Advance Praise for *Citi...*

"Engineers create many of the inventions that shape our society, and as such they play a vital role in determining how we live. This new book does an outstanding job of filling in the knowledge and perspective that engineers must have to be good citizens in areas ranging from the environment, to intellectual property, to ensuring the health of the innovation ecosystem that has done so much for modern society. This is exactly the sort of book that engineers and those who work with them should read and discuss over pizza, coffee, or some other suitable, discussion-provoking consumable."

—**John L. Hennessy**, president, Stanford University

"*Citizen Engineer* is the bible for the new era of socially responsible engineering. It's an era where, as the authors show, engineers don't just need to know more, they need to *be* more. The work is an inspiration, an exhortation, and a practical how-to guide. All engineers concerned with the impact of their work—and that should be *all* engineers—must read this book."

—**Hal Abelson**, professor of computer science and engineering, MIT

"Code is law. Finally, a map to responsible law making. This accessible and brilliant book should be required of every citizen, and especially, the new citizen lawmakers we call engineers."

—**Lawrence Lessig**, director, Safra Center for Ethics, Harvard University, and cofounder, Creative Commons

"Just as the atomic bomb brought us the citizen scientist, the computer has brought us the citizen engineer. This book is for engineers who take their societal responsibilities seriously, combining the idealism of dreamers with the pragmatism of builders."

—**Danny Hillis**, cofounder, Thinking Machines, Inc., and Applied Minds, Inc.

"In good economic times and bad, the forces driving companies to go green are getting stronger. Innovation will be the key to solving thorny environmental problems and creating lasting value for smart companies. Engineers are at the center of innovation. For businesses and the economy to experience the environmental and economic benefits of going green, we'll need engineers who read, understand, and act on the ideas in this book."

—**Andrew Winston**, author, *Green to Gold*

"The authors recognize the increasingly widespread impact of engineers on society in this new century and the resulting responsibilities that engineers now have. While engineering has long embraced safety in the designs of bridges and cars, not all of us consider the long-term environmental impact of our designs, or the importance of contributing to the knowledge base of engineering and honoring its intellectual property rights, as well as preserving the security and privacy of our fellow citizens who use our designs. I believe *Citizen Engineer* is a book that all of us teaching, studying, or practicing engineering should read, as well as those outside engineering who want to understand this force of change in the twenty-first century."

—**David Patterson,** professor of computer science,
University of California, Berkeley

"Douglas and Papadopoulos have created an essential road map for re-engineering products, services, companies, and commerce in ways that are environmentally responsible, economically profitable, and just plain elegant."

—**Joel Makower,** executive editor, GreenBiz.com;
author, *Strategies for the Green Economy*

"This book is the first to provide detailed guidance about eco-responsible product design and responsible use of intellectual property—two areas that are becoming vitally important to both the development of the engineer and the advancement of the engineering profession."

—**Dr. Bill Wulf,** professor of engineering and applied science,
University of Virginia; member, National Academy of Engineering

"With details and examples as well as principles, this book endows every engineer with a visceral connection to eco responsibility and to the new ways to create and use intellectual property."

—**Robert Sproull,** fellow and director, Sun Labs

"*Citizen Engineer* explains a critical transition of the engineering profession from technical focus to include social responsibilities and business context. This shift has changed the very nature of engineering as it is practiced today and as it must be taught in engineering degree programs."

—**Professor Steven D. Eppinger,** deputy dean,
MIT Sloan School of Management

Citizen Engineer

Citizen Engineer

David Douglas
Greg Papadopoulos
With John Boutelle

PRENTICE
HALL

Upper Saddle River, NJ • Boston • Indianapolis • San Francisco
New York • Toronto • Montreal • London • Munich • Paris • Madrid
Capetown • Sydney • Tokyo • Singapore • Mexico City

Library of Congress Cataloging-in-Publication Data
Douglas, Dave
  Citizen engineer : a handbook for socially responsible engineering /
Dave Douglas, Greg Papadopoulos; with John Boutelle.
      p. cm.
  Includes bibliographical references and index.
  ISBN-13: 978-0-13-714392-4 (pbk. : alk. paper)
  ISBN-10: 0-13-714392-3 (pbk. : alk. paper)
  1. Sustainable engineering. 2. Social responsibility of business. 3.
Engineering—Moral and ethical aspects. I. Papadopoulos, Gregory
Michael. II. Boutelle, John. III. Title.
  TA170.D68 2010
  620—dc22
                                        2009020712

ISBN-13: 978-0-13-714392-4
ISBN-10:   0-13-714392-3
Text printed in the United States with soy-based ink on recycled paper at R.R. Donnelley in Crawfordsville, Indiana.
First printing, August 2009

To our children, Ally, Cameron, Dana, Halley, Kathryn,
Jack, Madison, Michael, and Nicholas.

Contents

Preface

This book is a fusion of ideas, information, advice, and opinions from the authors, their colleagues, and dozens of other sources, brought together to provide you with the tools and insights you'll need to maximize your success in a new era of socially responsible engineering.

The information in these pages will be most relevant to engineers who design and build "things"—engineers in fields such as electronic/computer engineering, software engineering, mechanical engineering, materials engineering, automotive engineering, and so forth—although we believe engineers in all disciplines, managers of engineers, and even consumers will find useful information in this book. We've divided the book into four parts.

- **Part I: Advent of the Citizen Engineer** defines "Citizen Engineer," describes the trends that have led us to this new era of socially responsible engineering, and discusses what it all means—to engineers, to businesses, and to our society.

- **Part II: Environmental Responsibility** provides practical "how-to" information and resources to help you minimize the environmental impact of the products and services you're designing. It gives you an overview of what you need to *know*, things you need to *consider*, and what you need to *do* as you create ecologically and economically sound products, including (to name just a few topics)

    —Understanding and calculating the complete impact of a product or service

    —Defining strategies for key impacts such as greenhouse gas (GHG) emissions and water usage

—Trends in environmental regulations

—Whether "carbon neutrality" is sufficient as a business goal

- **Part III: Intellectual Responsibility** includes basic information about patents, copyrights, trademarks, trade secrets, nondisclosure agreements, standards, and licenses—and offers practical advice about how to maximize the economic opportunities intellectual property (IP) law presents while avoiding the potential pitfalls. For example, we discuss

  —The role of patents and when and how to file them

  —How to encourage other engineers to adopt and amplify your ideas

  —Pros and cons of various software licenses

  —Whether our system of IP controls maximizes innovation fairly

  —How to build communities to innovate and amplify your ideas

- **Part IV: Bringing It to Life** takes a look at some of the ways engineers—and engineering schools—are responding to the new realities and requirements of the new era, including

  —The growing momentum behind broader curricula in engineering schools

  —Advice for recent graduates and newly hired engineers

  —Examples of interesting projects with which Citizen Engineers are involved worldwide

You'll notice that the greatest emphasis falls on two broad topics that may not seem to be natural bedfellows: eco responsibility and intellectual property law. The reason is simple: These subjects have the greatest urgency to engineers today. They are redefining the way engineers do their jobs, yet most engineers are just beginning to understand the full impact each brings to bear on their work.

The book combines facts and viewpoints, and we've tried to be very clear about which is which. The subjects we discuss in each section can get enormously deep, so we've tried to give you enough basic understanding, along with pointers to further information, that you'll be able to continue exploring each topic. We hope you'll find the book useful in structuring your thinking and answering key questions.

Finally, a few notes about the book itself. Two of the key topics are environmental responsibility and intellectual property. Since the book will have physical manifestations and since it is, by definition, intellectual property, we've spent some time thinking about how this book lives up to the ideas it espouses.

First, let's look at the environmental impact of the book. If you're reading these words on a printed page, you're charmingly old-fashioned. This book is available in three forms, and only one of them is printed at all. We recognize the pleasures of reclining in a comfortable chair to read a book—but we also recognize the need to diminish the negative environmental impacts of traditional books. The publisher of this book, Pearson, has developed its own procedure to track wood back through the production process to the original forest, allowing the company to verify the sustainability of the papers it uses. Pearson also measures the carbon footprint relating to the shipping of its printed books around the world. The reuse/recycle rate for Pearson's unsold books and newspapers was 99% in 2007, in excess of the company's target of 95%. Pearson regularly reports on its progress to the United Nations as part of the company's commitment to the Global Compact.[1]

Our book is also available digitally. You can download it from a number of sources, and we hope you've taken the opportunity to acquaint yourself with the unique advantages of reading a book online: Digital versions are easy to scan; you can search for specific words or phrases; you can annotate and highlight electronically; and you can change the font size (those of us who are over age 40 appreciate this feature in particular).

Next, let's talk about the intellectual property that this book entails. By writing the book in the United States, we automatically get the privileges of copyright. In addition to the copyright, we have decided to license the content under a Creative Commons license, namely the Attribution-Noncommercial-Share Alike 3.0.[2] This means you are free to *share* this book (copy it, distribute it) and *remix* it to make derivative works under the conditions that your copies or remixes are for *noncommercial* purposes, that you provide proper *attribution,* and that you *share alike* any changes you make under the same (or a similar) license.

Finally, we have made every effort to properly recognize the works of others that we have leveraged in the writing of this book. If we've borrowed a line or a paragraph from someone's article or book, we've cited the source and referred you to the complete text. If we've used a resource such as Wikipedia to help us define a term or provide statistics that support our point, we have attempted to verify the accuracy of the content and cite the original source of the information.

One additional form of the book is interesting as it embodies both eco responsibility and intellectual property: We're making the book available as part of a living Web site (www.citizenengineer.org). We want you to do more than read the book; we want you to contribute to it. Add your thoughts about the new era of socially and environmentally responsible engineering. Insert your advice and lessons learned. Give the community tips for developing an environmental impact study. Got a better way to measure the carbon foot-print of a new device? Have some new information about an energy regulation? Let everyone know. This is a community effort; we welcome your participation.

Acknowledgments

Everything engineers create is an integration of many ideas. This book is no exception. Our aim was to produce something novel and useful by building on insights from many sources. We owe a debt of gratitude to the many people who have lent their time and energy to making this a better book. In particular, we would like to thank Jonathan Schwartz for his support of this project and Sun Microsystems for providing many of the examples used throughout the text.

Special thanks go to Al Riske for his hard work in the early days of this project. We would also like to express our deep appreciation to the following individuals who agreed to be interviewed, provide feedback about drafts, lend their expertise, or help us remedy errors along the way: Michael Anastasio, Kaj Arnöo, Subodh Bapat, Craig Carlson, Sheueling Chang, Michelle Dennedy, Greg Doench, Lori Duvall, Damien Eastwood, Steven Eppinger, Michael Falk, John L. Hennessey, Christy Confetti Higgins, Daniel Hillis, Mary Holzer, Joel Makower, Sohrab Modi, Dean Nelson, David Patterson, Mike Shapiro, Robert Sproull, Hal Stern, Harold Steudel, Dr. Ivan Sutherland, Michael Thurston, Chuck Vest, Andrew Winston, and Dr. Bill Wulf.

Thank you also to Carrie Motamedi, for her persistence and drive in keeping this project on track; and to Ingrid Van den Hoogen and Anil Gadre for their encouragement and enthusiasm.

Finally, we want to offer our thanks to our families for the inspiration and wondrous gifts they give us every day, and also offer up the following personal acknowledgments.

David Douglas: I would like to recognize the people who helped make engineering such an important part of my life. Steve Ward, Danny Hillis, and

Dave Patterson added to my technical education, but more importantly they taught me the fun and rewards of fully embracing a life of innovation. I'd also like to thank the eco team at Sun for being a constant source of inspiration. And finally, I'd like to thank my dad, Robert Douglas, who led the way and encouraged my technical learning from early on.

Greg Papadopoulos: I'm hoping that history will end up recognizing the essential work of Richard Stallman, the GNU Project, and the Free Software Foundation in pioneering the ideal of developer freedom. Richard, along with Lawrence Lessig, has had a fundamental influence on my views on the interplay between creativity and control. On the connection of engineering with society, much was catalyzed by my work with the Anita Borg Institute for Women and Technology, and conversations there with the late Richard Newton were particularly encouraging. But here, the real seeds were planted years ago by my father, Michael Papadopoulos, and my mentor, Michael Dertouzos: two Greek-Americans who reveled in that beautiful intersection between science and the human spirit. They both live in my memory every day.

About the Authors

David Douglas

Dave is senior vice president of cloud computing and chief sustainability officer at Sun Microsystems. He oversees the strategy and execution of environmental initiatives across the company, including enhancements to Sun's products in the areas of energy efficiency, cooling technologies, product recycling, and clean manufacturing. In addition, Dave is responsible for Sun's cloud computing business, with a focus on creating reliable, scalable, and sustainable computing and storage. He has been in the high-tech industry for more than two decades, including more than a decade of experience leading organizations to build more innovative, efficient, and eco-responsible products, and he has a long-standing passion about environmental issues. He earned bachelor's and master's degrees in electrical engineering and computer science from MIT. Dave sits on the board of the National Ecological Observatory Network (NEON) and is a senior fellow at the Breakthrough Institute. He currently lives in Concord, Massachusetts, with his family.

Greg Papadopoulos

With more than twenty years' experience in the technology industry, Greg Papadopoulos has held several executive positions, most recently serving as chief technology officer and executive vice president, Research and Development, at Sun Microsystems, Inc. He is responsible for managing Sun's technology decisions, global engineering architecture, and advanced development programs. He has also founded a number of his own companies, including co-founding Thinking Machines, where he led the design of the CM6 massively parallel supercomputer. Papadopolous was also an associate professor of electrical engineering and computer science at MIT, where he conducted research in scalable systems, multithreaded/data flow processor architecture, functional and declarative languages, and fault-tolerant computing. He holds a bachelor's degree in systems science from the University of California at San Diego, as well as master's and doctoral degrees in electrical engineering and computer science from MIT. Papadopoulos resides in Los Gatos, California, with his wife, Laurie, and has passions for cooking, wine, and eco-responsible living.

John Boutelle

A professional writer for more than twenty years, John has worked with and interviewed hundreds of engineers and executives from a diverse range of enterprises worldwide, including Adobe, Apple, Cisco, General Electric, Hitachi, Lam Research, Nokia, Novell, Oracle, Pacific Bell, Seiko, Sony, Sun Microsystems, VeriSign, and dozens of start-ups. Previously he was editor-in-chief of the *Orange County Business Journal* in Santa Ana, California. He holds a master's degree in business administration from the University of Michigan and a bachelor's degree from Pomona College. John resides with his family in Madison, Wisconsin.

Introduction
While You Were Busy Debugging...

Engineers have never been afraid of change. It's our job to effect change. We transform scientific principles and theorems into useful products and services of all kinds. In the process, we change the way people work, play, and live.

Virtually everything people touch today has been designed by engineers—from the cars we drive and the roads we drive them on to the mobile phones and GPS devices we use while we should be keeping our eyes on the road. And everything that is engineered is constantly evolving. The perpetual cycle of innovation, optimization, and exploration of new possibilities is what excites us. We relish the fact that the half-life of knowledge in our field is measured in months.

But now change has come to engineering itself. Over the past few years, while engineers have been busy slinging code or testing tolerances, the core requirements and responsibilities of engineering have been evolving faster than any underlying science or technology. And many of us have discovered that the new world of engineering is not the one we prepared for.

What has changed? While it's hard to put a finger on it, we all feel the effects. Some of us find that we're spending more time in meetings than in the lab. Or that we're working on an environmental impact analysis rather than a product design; or burning cycles trying to grok the nuances of the GNU General Public License (GPL) Version 3 as opposed to Version 2; or struggling to comply with yet another new data privacy mandate.

Taken together, these symptoms spell a seismic shift in what it means to be an engineer. Suddenly engineering is no longer solely concerned with finding a simple, elegant way to implement a set of design requirements. Success is no longer solely measured by the speed and efficiency with which design specs are met. Technological prowess and ingenuity are no longer enough; we need knowledge of subjects well beyond the scope of traditional engineering. A successful engineer needs to be part environmentalist, part intellectual property (IP) attorney, part MBA, and part diplomat—not to mention an expert in an engineering discipline, a great teammate, and a skilled communicator.

Recent trends are also redefining the role of the engineer in society. The increasing complexity of products leads to greater dependence upon engineering; yet most people don't understand engineering or the underlying sciences and technologies. This situation can be scary to the general public, and can lead to bad public policy and misconceptions that hold back new innovations. There is a pressing need for engineers to become more proactive with society—to engage, to communicate, and to lead.

This book takes a closer look at the nature of engineering today and provides practical guidance on topics of increasing interest and urgency to engineers, particularly environmental considerations of product design and intellectual property, licensing, and contractual considerations. The book also explores how eco-effective, techno-responsible products and services can translate to new opportunities for businesses and an accelerated career path for engineers.

In the course of writing this book, we talked to engineers, students, and researchers. We consulted with lawyers, environmentalists, administrators, and managers. We've blended their stories, experiences, and advice together with our own observations, all with a single, overarching goal: to help you become a more effective engineer, while maintaining every iota of the passion, visceral excitement, and creativity that drew you to this profession in the first place.

We're moving past the "Century of Science" into what we believe will be the "Century of Engineering." It's a period that will be both more exhilarating and more daunting for engineers than anything that has come before. It's an era that will redefine the way we think about ourselves as we continue to shape the way people interact with their world. It's an opportunity to become more socially responsible engineers and to create products that are more tightly aligned with our personal ethics. It's the age of the Citizen Engineer.

PART I
Advent of the Citizen Engineer

W e've entered a new era of engineering that is fundamentally changing the role of the engineer on the job and the engineer's relationship to society. In this part of the book, we'll share our evidence and opinions with you. But we'd also like you to consider what the new era means to you. Ask yourself how it might open new opportunities: for your professional development, your career aspirations, the breadth of your skills, and your ability to have a greater impact through your work. More broadly, consider what this new era might mean for your company and its prospects in the marketplace.

- How will it change the way your company innovates?

- How will it impact the way engineers collaborate—with each other and across the organization?

- How can engineers help society better understand the technologies and products that they create and work with?

- Should engineers become more influential and participative in public policy?

- Should engineers play a larger role in educating and shaping the public's view of technology and its implications?

While these are important questions to ponder, let's be clear: We didn't write this book simply to provide food for thought. After all, engineering itself is pure purpose—the application of knowledge to create something of value. We want to give you insights and information that you can apply to further your personal and professional development, your company's success, and the evolution of our chosen profession.

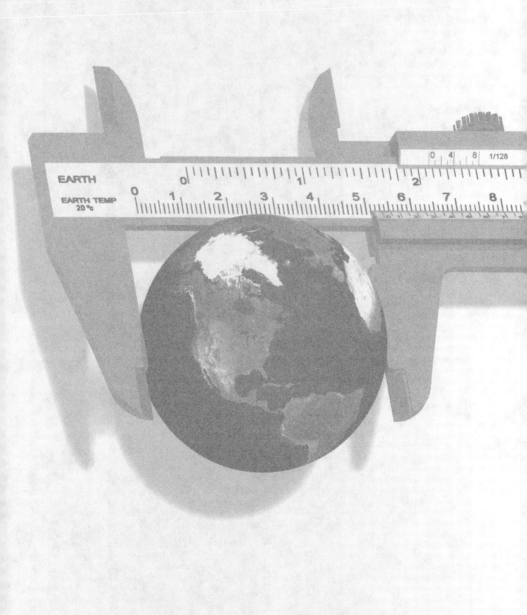

1

"Citizen Engineer" Defined

> "The scientist describes what is; the engineer creates what never was."
> —*Theodore von Karman*

To put it as simply as possible, Citizen Engineers are the connection point between science and society—between pure knowledge and how it is used. Citizen Engineers are techno-responsible, environmentally responsible, economically responsible, socially responsible participants in the engineering community.

You'll find additional shades of meaning in the term *Citizen Engineer* as you examine its elements separately. What is an engineer? An engineer is a constructive artist. The art of engineering is based on science and mathematics, where the tools and materials are technological. It's a constructive art because engineers build and optimize things. And yes, the intent here is to highlight the concept that engineering is *art*—interpretation, design, creation, invention, and expression—in contrast to the common stereotype that the profession is rigid and formulaic.

Laypeople often confuse engineers with scientists. For example, we're sure you've seen the incredible results from the Mars Rover missions trumpeted as a spectacular achievement of science, and a wonderful way to attract more students to math and science. As engineers, we look at the science components of the mission as interesting—but the engineering, ah, that was monumental! Hey kids, wouldn't you just love to *build* something that crawls around for years digging in the dirt of another planet? How cool is that?

Engineers are idealists, in every sense of the word. While we're focused on finding the simplest, most efficient solutions, we are also utopians. We find ourselves contemplating not only what could be better, but also what could *be*. We're drawn to visionary ideas and idealists. Over the years, engineers have demonstrated incredible courage in pursuing their visions—working for years on concepts that may or may not ever pan out, testing ideas no one else believes in. We also share the idealism of artists. We see beauty in new ideas, novel approaches, and new ways to tame complexity. We're a stubborn, meticulous, critical bunch, unforgiving of shoddy work and half-baked theories, abhorrent of conventional wisdom, but always eager to learn, to try, to push the limits, and to create something new.

At the same time, engineers are pragmatists. We must deal with the constraints of technological limitations, business requirements, and budget realities. We don't always have the time we'd like to implement our best ideas, or have access to the ideal tools or materials we'd love to use. Sometimes we must content ourselves with the knowledge that the world is better off with a *real* product that delivers 20% more energy efficiency today than with a *theoretical* product that might be 80% more efficient but is unlikely ever to make it to market.

Now consider what it means to be a citizen.

At one level, a citizen is simply a member of a community. But for most people, the term also includes a moral element. As a citizen, one has rights, certainly, but also responsibilities. Citizenship is more than participation in the community; it means working toward the betterment of the community through economic participation and public service to improve the state of all other citizens.

The blending of engineering and citizenship is nothing new. What's new is that engineers are being asked to extend their sphere of responsibility into new areas: the developing world, the environment, the proper use of intellectual property, security and privacy issues; the list goes on. At the same time, society is asking engineers to accept more responsibility for the impacts of the products and services they design. The world is not blaming engineers for climate change, loss of data privacy, and so on, but society is making it overwhelmingly clear that since engineers had a role in creating these challenges, engineers must accept their role in addressing them as well.

We're not saying that engineers should be wracked with guilt if their next product has a nonzero carbon footprint, uses a scarce natural resource, or offers only a modest environmental improvement over a previous design. Engineers must be pragmatists. Your product can never be ideal. Instead, what we are saying is that you need to be aware of the impact your product has—and be creative about minimizing its negative impacts. We're asking you to consider new options and explore new possibilities.

In a sense, "Citizen Engineer" is an ideal. There is no formula or set path for becoming a Citizen Engineer, no specific set of attributes that all Citizen Engineers possess, and no certification process. Attainment of

Citizen Engineer status is deeply personal; only you can define the requirements given your situation, and only you can measure your progress toward the goal.

It's interesting to note that many of the youngest members of the engineering community are helping to drive the new era of socially responsible engineering. Typically, the responsibilities of citizenship are something one grows into as one ages and accumulates wisdom and perspective. What's the explanation? Certainly part of it is that the young have the most to worry about when it comes to the future of our planet. But we believe it's more than that.

We believe the new generation of engineers understands the unprecedented opportunities before them—to create better products and to make a positive, lasting impact on our planet. They can't wait to get started.

Responsibilities of the Citizen Engineer

If you're an engineer you've already enjoyed many rights and privileges. You've received an excellent education. You've put yourself in a good position to effect change and exert your influence on society. And you get paid to innovate and create—who could want more than that?

But with these privileges come responsibilities.

First, the basics: Engineers have an ethical obligation to make decisions that are consistent with the safety, health, and welfare of the public, and to disclose factors that might endanger the public or the environment. Our engineering community has met this obligation exceedingly well over the years. When people cross a bridge or board an airplane or get into an elevator, they're not wondering whether the engineers who built it were socially responsible. Society's trust for engineers is deeply ingrained. And today, society trusts us to get these new things right as well—from the next genetically engineered foods or cancer treatments, to the nanotechnology underlying new ways to create clean water, to the software in your hybrid car that ensures that the brake pedal really does stop the vehicle.

But Citizen Engineers are moving beyond the basics, and are upholding even higher ideals. They are more than problem solvers. They are speaking out on issues within their realm of expertise and engaging in the political process. They're staying abreast of issues that impact their field and are helping to educate others who may be impacted by developments in their field. And they're embracing new forms of responsibility that are becoming increasingly important to our society as a whole.

- **Environmental responsibility:** Citizen Engineers have both the opportunity and the moral obligation to consider the total environmental impact of the products and services they design—over the entire life-cycle, from raw materials through manufacture, assembly, distribution, sales and marketing, use, recycling, and disposal.

- **Techno responsibility:** Citizen Engineers use the technological innovations of others responsibly and ethically, building and contributing to the knowledge base in their field. This includes the responsible use of intellectual property, adherence to the terms of licenses, ethical handling of data about others, and honoring the terms and the intent of nondisclosure agreements and other contracts relating to use of ideas.

- **Customer/stakeholder responsibility:** Citizen Engineers are responsible for ensuring the safety, security, and privacy of the people who buy and use the products they design. They must also consider issues such as accessibility; they're responsible for testing, upgrading, and improving their products over time; and they must account for unintended consequences of their products. Citizen Engineers also work closely with customers to ensure that they are partners in minimizing the environmental impact of products and services. After all, customers often control the project requirements, the time frame, and the budget that engineers are working under.

Knowledge Base of the Citizen Engineer

> "It's not what you know, it's who you know."
> —*Old adage*

> "It's not who you know, it's what you know."
> —*Engineering manager*

> "It's not just what or who you know, it's who you are."
> —*Citizen Engineer*

To be successful in the new era of engineering you're going to need to be more than an engineer. That's true whether you're a student, a new hire at a company, an engineering manager, an entrepreneur, or a seasoned engineer with 30 years of experience.

New requirements are encroaching on the traditional tasks of engineering. Most design specs will soon reflect a requirement to minimize environmental impact—do you know what to consider and how to comply? Many new software projects may consider using open source software—do you fully understand the nuances of the various licenses?

In addition, being a Citizen Engineer means being a leader. And your ability to lead—not to mention your influence and impact as an engineer—will increasingly depend on your ability to communicate, collaborate, and participate across the organization, not just within the engineering department.

As a society, we'll need engineers to take the lead in solving many of the most pressing challenges we face: the environment and climate change, increasing data security and privacy issues, and the ability to sustain quality of life without depleting natural resources. We'll need lawyers, environmental scientists, business experts, and other professionals too, but engineers are critical because only engineers have the expertise and specialized knowledge required to effect these types of changes.

To put it simply, you won't need to "know" more than a traditional engineer; you'll need to "be" more. You'll need to play multiple roles, possess broad knowledge in a variety of disciplines, and know when to seek professional assistance. Here are a few key areas of expertise you'll need as a Citizen Engineer.

Technology

First and foremost, you have to know your stuff. That's getting more and more challenging as fields of technology and engineering become increasingly specialized, but for most engineers it's a labor of love. You keep learning because you're compelled by your own curiosity and interest. If staying in sync with new developments in your field is burdensome, that's telling you something.

Ecology

Within the next few years, virtually every new product or service will include environmental considerations as part of its core design specifications. Whether you see environmentalists as saviors, anti-business tree huggers, or something in between, you will need to be able to respond to the issues, laws, and new requirements for eco-responsible products and services—and be able to take advantage of the opportunities that eco-responsible products represent.

Intellectual Property

Depending on your attitude and experiences, the world of intellectual property (patents, licenses, contracts, etc.) can be a blessing or a curse. But with engineers collaborating in groups across the Internet, widespread use of open-source software, increases in data collection by products and companies, and more products and services with a large digital component, an understanding of intellectual property (IP) is now central to any engineering project. You need to know how to protect your own ideas through patents, copyrights, trademarks, and so on—and you need to protect the security and privacy of customer data as well.

Business

For more than a few engineers, part of the appeal of engineering is insulation from the harsh realities of the business world. The vision of working in a corporate lab, sealed off from distractions, free at last to pursue visionary ideas unfettered by such inane activities as cost justification or customer meetings, can be a powerful draw to some. Unfortunately, that vision has never matched reality, and engineering increasingly requires more direct involvement with every aspect of business, including finance, sales and marketing, channels, customer support, and competitive analysis. Here's the upside for die-hard techno purists: The more you know about business, the more control you'll have over your ideas and innovations. If you can clearly articulate how your project will solve problems for customers and improve the company's bottom line, you will increase not only the likelihood of funding for your project, but also your influence within the organization.

Public Policy

Whether you're aware of it or not, your products embody your ideals. And as a result, engineers are activists, whether they want to be or not. But engineers are also beginning to realize how important it can be to educate those who make public policy decisions about technology. Nuclear power and genetically engineered foods provide cautionary tales about what can happen when politicians and the public don't understand the realities behind new innovations. Time and again the public has heralded a technological advance as a savior, only to demonize it months later, when in fact both views were way too extreme. Citizen Engineers need to play a more active and assertive role in providing that education. And conversely, Citizen Engineers need to learn more about how public policy actually works, and to be educated and informed about the process itself. It is our duty to make the effort to have the dialog with our communities, not the other way around.

Collaboration

In discussing this book with dozens of engineers, there wasn't much every-one agreed on—but here's one thing: Collaboration skills are becoming every bit as important as technical knowledge for engineers. In short, you have to be a good teammate. Even among NASA astronauts, team skills are now seen as more vital than piloting skills.[1] It's no different in the close quarters of the corporate lab, conference room, or networked workspace. Working together, especially in small, highly communicative teams, is what works.

As you read through the previous sections, a question may have crossed your mind: How am I supposed to do all this? Chances are you're already under intense pressure trying to meet project deadlines, keep pace with new developments in your field, and balance your work and private time.

We're not advocating that you become an expert at everything that touches your work. But you'll need to know the basics about each of the pre-ceding topics, and you need to know when to reach out to other profession-als who can help you with specific issues. That's what this book is all about. We want to help you use your time most efficiently as you come up to speed on new areas of knowledge that will impact your job, your project, and your success as an engineer.

2

How Engineering Got Its Paradigm Shifted

The new era of engineering is being shaped by forces that have been building over several decades as well as by more recent events. This chapter looks at the trends that are reshaping and redefining engineering as we knew it.

Changes in the Nature of Engineering

Much of the transformation in engineering is being driven from the inside out. In this section, we'll explore changes in the scale of engineering, the increasing levels of collaboration among engineers, and the upward spiral of engineering influence on everything from products to corporate profits.

Engineering on a Whole New Scale

We're seeing radical changes in the scale of engineering on multiple fronts. First, the diversity and complexity of materials used in new products today are staggering, and some of the synthetic materials and chemical compounds we're using are creating issues we didn't anticipate and are still struggling to deal with: how to dispose of nonbiodegradable materials; how to prevent toxic and hazardous substances from leaching into our air, water, and soil; and how to reuse materials that were designed for very specific purposes.

Second, we now have the ability to produce things by the billions, the capacity to distribute them globally, and the markets to consume that kind of scale. Global markets, global fashions, and global consumer trends result in mass production of successful products—*and the repercussions of success can far exceed anything engineers originally envisioned.* The gizmo that's on your drawing board today could be on tens of thousands of store shelves in an incredibly short time. So, any unresolved environmental issue, design issue, or intellectual property (IP) issue gets proportionally amplified, often globally.

Simply put, the problem of scale can transform today's solution into tomorrow's problem. Did Henry Ford have any inkling that a hundred years after the introduction of the Model T we'd be trading streets full of manure for global climate change, due in part to the exhaust of more than a billion cars?

"When you scale things up, you change the fundamentals, and engineers are called upon to go back and innovate again and make the product or service sustainable," says Sun engineer Subodh Bapat. "We've seen this happen time and again—products that solved one problem and created another because of unforeseen scale issues: Teflon; DDT; more recently, CFLs [compact fluorescent lamps; see Figure 2-1] with their mercury content. I think today's Citizen Engineers will need to start thinking further ahead and consider the potential impact of success—from a sociological perspective. Our challenge will be to design products that can succeed in the marketplace without damaging our environment or our culture."

FIGURE 2-1   Examples of CFLs

That's no simple task, particularly given the accelerating pace of change in everything from technology to geopolitics. "Perhaps it is always hard to see the bigger impact while you are in the vortex of a change," wrote Sun cofounder Bill Joy in *Wired* magazine.[1] "Failing to understand the consequences of our inventions while we are in the rapture of discovery and innovation seems to be a common fault of scientists and technologists; we have long been driven by the overarching desire to know what is the nature of science's quest, not stopping to notice that the progress to newer and more powerful technologies can take on a life of its own."

While product volumes can scale to huge sizes, engineers are using building blocks of ever-smaller scale. We've gone from working with tons of raw materials to transistors, nanomachines, photons, and base pairs. Today we're working at an atomic scale to create new raw materials that can be used to create lighter-weight, sturdier autos and aircraft; we're manipulating subcellular structures such as recombinant DNA to create new biologics, pharmaceutical products, and potentially even new species; we're using microorganisms to develop everything from new breeds of biofuels to a new generation of fermented beverages. And as the size of our source material continues to shrink, the opportunity for building new and better products continues to grow.

Consider also the changing scale of our computational capabilities and how they have radically altered what it is possible for engineers to do. One eminent engineer at Sun put it this way: "What's different about today's era of engineering comes down to two things: mathematics and computers. Go back a hundred years or so: Thomas Edison tinkered stuff into existence. He tried thousands of filaments for his incandescent bulb—one by one. And he used direct current because when you're dealing with alternating current you have to work with complex numbers, and he simply didn't have the math to do it. Westinghouse, on the other hand, had Nikola Tesla and Charles Steinmetz, who could handle the math involved. Today, the ability to use high-powered computers to do mathematical modeling and simulations is absolutely vital and central to the nature of engineering. And as the scale of computing increases, so does the power and influence wielded by engineers."

A final aspect of scale to consider is the temporal impact of the products and services we design. "The scale of our impact on the environment now has effects on time scales much longer than those of typical human time horizons," wrote Kirk R. Smith, professor of Global Environmental Health at the University of California, Berkeley, in an editorial about climate change.[2] "The archetypal example is perhaps global pollution leading to global climate change. We are now well into a planetary experiment on the effect of injecting a bolus of warming pollutants, three to four times natural levels, during an instant of geological time. Nothing much happens at first, but analysts say that much more is set to happen unless we mend our ways soon."

Simply put, Citizen Engineers will need to consider the long-term issues of what they produce—air pollution, climate-changing emissions, potential public health issues, and so on—at the same time they're addressing today's problems.

In summary, we can now design things that are more complex, can last longer, have more impact, and can be replicated more times than anything that came before it. This gives tremendous power to engineers, but also tremendous responsibility.

Pervasive Collaboration

The Internet also broadens the scale of engineering collaboration. We now have instant access to the thoughts, insights, and feedback of other engineers, community members, partners, suppliers, and customers, all the time. The result is a surge in the sheer volume of new ideas (because more ideas from more sources are instantly accessible); constant growth in the size of engineering teams; and the demise of the mythical "lone-wolf inventor" as an icon of engineering.

"People had this image of a mad scientist working alone in a lab, and you slide a pizza under the door every once in a while until they emerge with this ingenious new thing," says Sun's Bapat. "And it's so laughable. Engineering success today is about teamwork; teamwork is about communication skills and collaboration tools; and the network is the ultimate collaboration tool."

"Individuals have more tools and more power as engineers than ever before to both receive information and communicate with others," says Sun engineer Mike Shapiro. "And the individuals who work most effectively are the ones who seek out information and input from their colleagues."

For individual engineers, the Internet means instant access to the knowledge and research of colleagues around the globe, an incredibly powerful tool. For groups of engineers, it means collaboration can take whole new forms. For example, development work on Sun's Solaris operating system is now perpetual: Software engineers in the United States work all day on new features, enhancements, and bug fixes, then transfer their work over the network to another team in Asia, which works all day and then transfers its work to another team in Europe—so that 24 hours a day, someone somewhere is working on Solaris.

The Internet has also enabled competition among engineering teams—even teams from the same company working in different parts of the world. At GM, for example, two design teams more than 6,000 miles apart—one in Detroit and one in China—competed for the right to reengineer the Buick LaCrosse for the Chinese market.[3] After an intense internal competition, the Chinese team won complete authority over the interior design, while the exterior was handled in the United States, but with a great deal of input from China. Eventually the two teams collaborated, harnessing the network to share ideas and design concepts.

The network also makes it possible for ad hoc communities of engineers to spring up overnight to work on specific technologies or technical issues. Some projects spring up as communities of like-minded individuals. In other cases, software projects initiated by corporations are virtually taken over by these communities—to the benefit of the corporation, the community, and consumers alike. We could write a separate book on that topic. At Sun alone we've seen more than a hundred communities form to build on our core technologies, from performance tuning Open Solaris to improving Java security to translating products and documentation to additional languages.

Broader Influence for Engineers

As the scale of engineering increases, so does the influence wielded by engineers. You see the evidence all around you—literally. You're surrounded by things engineers created, and chances are you're using more "things" than ever before.

You've probably also noticed the increasing clout of the engineering department within the corporation and the growing number of engineers in the executive suite. You personally may be benefiting from the premium salaries paid to top engineering talent and the fierce recruitment battles for the best and brightest from our universities (not too different from the competition for running backs and defensive ends in pro football). But there are other dimensions to this widening sphere of influence.

Salaries for engineers are rising—and rising faster than those of other professions. According to recent data from The Engineering Income & Salary Survey, median salaries for engineers are up more than 10% from 2006, and up more than 19% from 2005. The average starting salary for an engineer with a bachelor's (four- to five-year) degree ranges from $36,000 to $50,000—significantly higher than salaries for graduates with bachelor's degrees in many other fields. In comparison, lawyers starting out after at least seven years of school average just $45,000.

Higher economic impact and higher salaries translate directly to greater influence within the organization. One interesting piece of evidence: The most common undergraduate degree for CEOs of Fortune 500 companies is not marketing, sales, or finance—it's engineering, with 20% of all CEOs holding engineering degrees.[4]

But it isn't just their growing economic impact that has increased engineers' influence. Once again the network plays a crucial role. Sun engineer Mike Shapiro provides an excellent example: "When I was in college I wrote a program for the Macintosh—a little utility, which I gave to friends and put on an FTP site. With it I included a message: something to the effect of 'this is free, but if you like it, send me a postcard.' The utility became very popular, and I literally started getting pounds of mail from around the world. People sent me everything from a ten-lira note to pictures of their children—even recipes. That's when I realized how much influence I can have as an engineer, and how much reach my ideas can have. And the fact that you can have that level of influence that quickly—it's very powerful and revolutionary."

As this example shows, engineers are now often getting closer to their users, using the network to stay in touch with them over time. Millions of engineers blog and "tweet," including thousands at Sun. Many software companies urge their engineers to be active on the company's online forums, interacting directly with customers on their issues and ideas. And in almost every case we've seen, engineers have accepted the responsibilities that come with this unfettered access to the "real world." As a result, companies are pushing engineers to the edge of the company, not burying them inside for fear of what they may say.

Another aspect of engineering influence is subtler but equally important: the ability to change the way people think. And you don't have to work for a large organization or an economic powerhouse to do that.

Consider the example of Tesla Motors, the Silicon Valley start-up company that's building the first high-performance, consumer-oriented electric cars. The company's first production car, the Tesla Roadster, has a range of 221 miles (394 km), accelerates from 0 to 60 mph (100 km/h) in less than 4 seconds, and has power costs of only about 2 cents per mile. It emits no exhaust. Simply put, it's the car that proves that an eco-effective yet practical vehicle is technologically within our grasp. The company's financial future, however, is in question. Despite having raised more than $100 million, it remains to be seen whether Tesla's business model proves viable.

The point is, whether or not Tesla Motors survives, it has already won a huge victory in terms of influence. "I would argue that Tesla has already changed the world," says Craig Carlson, lead software engineer at Tesla Motors. "If we have to close our doors tomorrow, we will still have had a major impact, because people have seen that it is possible to build a very impressive electric vehicle now, and that wasn't clear a few years ago. So, I'd argue that we've been very influential even if the road isn't full of Tesla vehicles."

Combined with the disintermediation we've described, this ability to change minds and alter purchase decisions is evident in many forms today: blogs from engineers offering tips on minimizing power consumption in new devices; YouTube videos showing new ways to use photovoltaic cells. Increasingly, engineers have both an important message and the medium to deliver that message.

When you combine the increased economic impact engineers have; the increasing speed, power, reach, and capabilities of the network; and the growing ability of engineers to shape attitudes and purchasing decisions, the result is an enormous expansion of influence for engineering.

Externally Driven Changes in Engineering

We've said that the emergence of the Citizen Engineer is partly the result of trends within engineering itself and partly due to new pressures from society. Let's take a look at a few of the society-driven changes.

The Green Explosion

Flip through any newspaper or magazine and you'll see that as a society we've become obsessed with eco-friendliness. The News section will have articles about the opening of a local recycling center or a new battle between the EPA and the state of California over the jurisdiction of eco legislation. The Sports section will include articles about new eco-friendly packaging for hydration products and more energy-efficient lighting at stadiums. The Style section will discuss the latest green fashions featuring clothing made from bamboo or hemp. The Home section will provide pointers about cutting the power consumption of appliances or making the move to geothermal heat-transfer systems. And the Business section will be filled with articles about new corporate environmental initiatives, the latest take-back legislation, or sales projections for the Smart Car.

This wave of green isn't just a passing fad. Legitimate concerns over climate change driven by increased production of greenhouse gases (GHGs), shortages of fresh water, and issues with disposal of complex chemicals are just the tip of the proverbial iceberg (oh yeah, those might be disappearing also). In addition, the increasing scarcity of energy sources and critical natural resources will have growing economic impacts. Jeffrey Immelt, CEO of GE, says "Green is green," recognizing the tight coupling of our ability to produce sustainable products and services, and our ability to have a growing economy.

In response, the new generation of green products and gizmos has arrived in full force. On display recently at the Consumer Electronics Show in Las Vegas:

- Solar-powered backpacks that can do everything from charging your cell phone to heating water for an outdoor shower

- A laptop with a plastic case made from corn rather than petroleum products

- Smart power adapters that don't waste as much electricity

- New silver-zinc batteries that could replace the highly toxic lithium-ion batteries that power most cell phones and laptops today

- Home automation systems that can be set to make intelligent decisions about when to turn themselves off

What it all means to Citizen Engineers is that the market for eco-effective products is here; consumer demand for eco-effective products is real; and efforts that ignore standard practices for eco-effective design will be punished by the marketplace.

Corporate Social Responsibility

Corporations have a major environmental impact, but they can have a huge effect on society as well. Companies touch the lives of their employees in some way almost every day, and they impact everyone they do business with as both a buyer and a seller of goods and services.

In recognition of this, the practice of Corporate Social Responsibility (CSR) is exploding. Shareholders are asking whether the companies they invest in are exposed to unnecessary risks or legal liabilities. Companies are checking into the ethical behavior of their major suppliers. Watchdog organizations are keeping tabs on every direct and indirect impact of corporate behavior. And consumers are reacting with a vengeance when bad news comes in any form.

For engineers, this means a new set of things to pay attention to. Selecting a part that is made by only one company of questionable background is probably not a good idea. Outsourcing to factories with bad pollution records will be noticed. And the discovery of a competitor's files on your company's computers can be expensive on many different levels.

Security and Privacy Concerns

Virtually everyone has a creepy story involving an online scam, identity theft, or fraudulent use of sensitive information. Maybe you returned home from a vacation and your credit card statement included the purchase of a new plasma TV that you don't recall buying. Maybe you got a couple hundred bounce-back emails one day because your computer or email address has been usurped by some kind of "zombie" or "bot" to send spam. Or maybe you responded to an email about a recent purchase on eBay and noticed that the Web site you were directed to was a fraudulent subdomain (ebay.cbay.com or www.paypai.com; see Figure 2–2).

FIGURE 2–2   Examples of Authentic-Looking Brand Marks Designed to Entice Users to Fraudulent Web Sites

We all know that malware, viruses, phishing scams, man-in-the-middle attacks, worms, logic bombs, and various other exploits have been with us for many years and will probably always be part of the online landscape. But few of us realize just how serious this problem is becoming.

In 2007, the losses from phishing attacks in the United States alone reached the multibillion-dollar level worldwide, according to the Anti-Phishing Working Group (APWG). And recent exploits underscore the growing sophistication and creativity of hackers. It's not just the naïve who are victimized.

In 2007, for example, a wave of "pump-and-dump" scams on stocks trading on NASDAQ victimized seven online firms, earning the perpetrators $732,941 in illegal profits.[5] The attackers used phishing techniques to break into the accounts of innocent brokerage customers. They sold off holdings and used the money to purchase penny stocks, then used email promotions to artificially inflate the price of those stocks and sold them at a profit. Another example: An engineer from Virginia (an engineer!) thought hackers who broke into his computer stole only his bank account information; but a hidden virus continued to record his every keystroke, and hackers soon got his new account information and passwords too.

There is even a thriving underground economy for stolen personal information. A complete identity—a package of a person's name, date of birth,

Social Security number, and credit card and bank account numbers—sells for about $14 to $18, according to Alfred Huger, vice president of Symantec Security Response.[6] Huger says a verified PayPal account sells for $50 to $500, depending on the available balance in the account. And there is a large and growing market for malware, the software that hackers use for phishing and other exploits. For subscriptions starting at about $20 per month, organized hacking gangs sell fully managed exploit engines that spyware distributors and spammers can use to infiltrate systems worldwide.[7]

And the threat to our security and privacy doesn't come only from malicious exploits. Many legitimate companies collect massive amounts of data about us: the time, date, and particulars about our past purchases; the frequency of our visits to their stores or branches or Web sites; and the tendencies, tastes, and preferences we've displayed about everything from reading materials to frozen foods. We trust them to protect our privacy, but we have no real assurance that they will, beyond the "privacy protection policy" that we scroll by on our way to the Checkout button. Nor are we even aware, necessarily, when companies breach our trust and sell or exploit our private information.

Some of the most obvious and egregious problems with privacy protection are now so commonplace that we barely notice them. Even today, when you log onto Amazon.com, for example, you'll be greeted with a message such as "Hello, Greg Papadopoulos. We have recommendations for you." Right next to that you'll see a link that says "Not Greg?" It might as well say "Not Greg but want to use Greg's account anyway?" or "Not Greg but interested in Greg's personal information?" Identity management has a long way to go.

What all this means to engineers is that there will be increasing pressure—and incentive—to design products and services that ensure data security and protect consumer privacy while still allowing companies to use data to improve customer service. And beyond this, a Citizen Engineer needs to be active across the entire dialog of product and service design and planning—to be informed and to reason through social consequences. It is, after all, systems designed by engineers that have at once created the opportunities and the potential for misuse. We can't enjoy the benefits without also assuming the responsibilities.

Rise of Digital Goods

Our notion of a "product" as a physical entity has been challenged by the rise of digital downloads. Huge numbers of consumers now have the tools to create, market, sell, and collect money for movies, photography, music, games, and more.

Perhaps not surprisingly, the environmental pluses and minuses of digital goods create some difficult questions for engineers. How will you encourage people to purchase digital goods and make it more convenient for them to do so while protecting the rights of the people who created the digital content? What is the environmental impact of digital products and how will you measure it? Which IP laws apply to digital goods and how will you stay abreast of fast-changing IP laws relating to digital content?

Digital rights management (DRM) systems are available to help control the distribution of protected content, but these systems raise additional questions. Does DRM quash creativity and freedom of expression? After all, in both science and art, innovators build on each other's work. And is tighter control of digital goods a good thing for business in the long run? Won't consumers find workarounds, as they did with Napster, and create a perpetual game of cat-and-mouse with businesses and content owners?

It's up to all of us who see ourselves as Citizen Engineers to advocate policies and to design technologies that respect the legitimate needs and current rights of honest users.

New Laws, Tighter Controls

Engineers who were once preoccupied by Moore's law are now dealing with more laws—an ever-increasing set of regulations and controls governing their work. The world is increasingly a patchwork of laws that regulate everything from chemical content to noise levels to recycling to safety labels.

In addition to laws, engineers often face tighter controls and processes within their own companies. These often come in the form of formalized quality assurance methodologies. Six Sigma, Statistical Process Control (SPC), Total Quality Management (TQM), the Shewhart Cycle (a.k.a. Plan-Do-Check-Act or PDCA), Top Quality Control (TQC), Zero Defects, and Software Quality Assurance (SQA) are just a few examples.

Engineers understand that process and constraints are important, but they almost always make engineering more complex. In addition, there's a very fine line between a well-designed process that adds value with limited overhead and a poorly designed process that inhibits engineering and provides little or no value add.

From an engineering perspective, not a lot can be done. It is important to educate our legislators and corporate process wonks on what will be effective and what will be just plain expensive. But it's also important to invest the time to understand the laws and processes that do exist, in order to build them into your designs and activities so that you can stay legal both inside and outside your company.

Perspectives on an Engineering Transformation

In the course of writing this book, we've had an opportunity to speak with many of our colleagues about the emerging era of socially responsible engineering, and the different perspectives we've heard are fascinating and illuminating. We'll end this part of the book with a few viewpoints that lend additional contour.

First, as several of our colleagues have pointed out, what's "new" about this era is not the focus on responsibility; engineers have always taken their civic and moral responsibilities very seriously. What's new is the groundswell of interest among engineers in extending that same conscientiousness to new areas: to the environment, to the proper use of intellectual property, and to public policy, for example.

Nor is it new that engineers now have to "juggle many balls" and broaden their knowledge beyond the field of engineering. But our colleagues sense that what we're seeing today is a growing recognition of the importance of "breadth" in the face of steadily mounting pressure to narrow one's focus.

It's a very interesting phenomenon. Today, as Sun's cofounder and chairman, Scott McNealy, once said, "Technology has the half-life of a banana." As technologies and fields of engineering continue to grow more specialized, more and more time and effort are required to master any engineering specialty and stay abreast of developments. Yet the new generation of Citizen Engineers has recognized that if you only know your stuff, you'll eventually limit your role, your professional options, and your influence. And in ever-growing numbers, they're looking to broaden their formal education and extend their skill sets beyond traditional fields of engineering.

Several engineers have voiced the opinion that engineering itself isn't changing at all—economics are. "Society keeps changing, so the economics of meeting society's needs keep changing," one of Sun's most distinguished engineers commented. "Engineering hasn't changed; it is still the art of making something useful that's also economic. And if you argue that social responsibility boils down to economics, then what we're seeing today certainly isn't anything new."

There are plenty of examples of how social responsibility drives economics and how that in turn drives innovation. However, we think social responsibility is about more than economics, and that being a Citizen Engineer is about more than filling market needs. Engineers do not merely cater to the tastes and demands of consumers; they also influence and guide the attitudes and preferences of a society and create new capabilities. When Alexander

Graham Bell demonstrated the telephone to President Rutherford B. Hayes in the White House in 1876, there was no consumer demand for a new communication device. In fact, President Hayes remarked, "That's an amazing invention, but who would ever want to use one of them?" The telephone was a breakthrough innovation that created a new market and changed society.

Finally, a number of people have pointed out that while there is indeed a shift toward more socially responsible engineering, it's actually part of a broader trend that's gaining momentum in relatively affluent countries.

"There is a growing realization that we can afford to be eco-responsible and use our influence and wealth not only to do good for ourselves, but also to help the developing world," says Steven Eppinger, deputy dean at the MIT Sloan School of Management. "It's within our ability and it is part of our shared duty. I've seen a tremendous amount of enthusiasm and activism around that idea—not just at the engineering school here at MIT, but on campuses across the country."

We're glad to see it happening, and we hope this book will be a catalyst for even greater momentum behind the movement toward environmental and techno responsibility.

Part I Summary, and What's Next

This part of the book has discussed the nature of the change in engineering and what it means to engineers; now let's turn our attention to the core topics that comprise the "handbook" portion of the book: how to make products and services that are more environmentally responsible, and how to use ideas—your own and those of others—more responsibly as you create the new generation of products and services.

While this part of the book took a philosophical bent, the next parts get much more prescriptive, helping you sort through the complexities of assessing environmental impact and dealing with IP law. We'll give you advice based on our experience, the input of our colleagues, and expertise from many other sources, and we'll give you pointers to additional information you may find useful.

We hope this part of the book has given you some context for understanding the era of the Citizen Engineer and the new requirements you'll face. Now let's roll up our sleeves.

Environmental Responsibility

In many ways, environmentally responsible design is still in its infancy, yet it is already a complex, multilayered subject that can be daunting for engineers. It can be next to impossible to identify and measure the full range of environmental impacts of a single product or service, let alone determine priorities for improving environmental performance. The issues you have to deal with include

- Determining the carbon footprint of a product or service

- Understanding the impact of different sources of electricity

- Knowing which chemicals and materials are desirable—and which should be avoided

- Maximizing the recyclability and minimizing the waste of a product

- Determining the fresh-water footprint of a product or service

In this part of the book, we'll use a two-part approach to tackle this large and complex design challenge. First, we'll break down the problem using a basic lifecycle model, focusing on how a product or service is made, used, and renewed. With this model we can examine the environmental impact of materials extraction, manufacturing, supply chain and employees, packaging, operation, recycling, take-back, and more.

However, *our approach does not require you to analyze and act on every conceivable variable that may relate to the lifecycle model.* Instead, as the second part of our approach, we'll present a framework to prioritize your actions—from determining the areas of greatest potential impact to identifying the low-hanging fruit for simple improvements.

We believe this approach will help you increase the eco-effectiveness[1] of your products quickly and see opportunities you might otherwise have missed—not just for creating better products, but also for improving business results through environmental responsibility. And while this approach works for individual products, it can also be used for product lines, or, with small changes, whole companies. In fact, we use a variation of this approach as we work to decrease the overall impact of Sun Microsystems as a company, and we have found that it accommodates many different scenarios.

This topic is very broad, and the specific set of challenges will vary from one situation to another. Therefore, the intent is to provide you with a starting point—a way to reason through the complexity and find the best approach for your specific situation.

Let's begin with a closer look at some of the variables that go into "environmental impact" and a discussion of what we, as engineers, can do to begin to minimize these impacts.

3

Environmental Impact: The Big Picture

The planet's population is now approaching 7 billion—an increase of about 5 billion people in just the past five decades—and the total population is likely to increase by another 1 billion people in the next decade. Analysts now expect that the ranks of the middle class (people who may want your products!) will swell by as many as 1.8 billion in the next 12 years.[1]

You've probably seen similar projections, and even though you know intellectually that an extra couple of billion people represents a sustainability challenge, it can be hard to relate those huge numbers to your job. So, to make the scale more real, let's work through what it would mean to give the next 1 billion middle-class citizens of the world a single 60-watt incandescent light bulb.

Each bulb weighs about 0.7 ounce, including the packaging, so a billion of them weigh around 20,000 metric tons, or about the same as 15,000 Toyota Prius cars. As an engineer, you know that multiplying anything by 10^9 makes a big number, but even from this simple case you start to get a feel for how dramatic the scale is in real-world terms.

Next, let's turn on those light bulbs. If they're all on at the same time, they would consume 60,000 megawatts of electricity—and that would require 120 new 500-megawatt power plants to keep them burning. Luckily, our imaginary middle-class consumers will use their light bulbs only four hours per day, so we're down to 10,000 megawatts at any given moment. However, that means we'll still need 20 new 500-megawatt power plants. If coal-fired, each of those plants burns 1.43 million tons of coal per year.[2]

That doesn't sound like a good idea from an eco perspective, so let's try solar power for our light bulbs. If we use current commercially available solar technology, we'll need roughly 50 square kilometers of solar panels, or more than one-third the land area of either San Francisco or Boston. Hmmm. So, let's try wind power instead... We'll still need one-tenth of all the wind power produced in the world in 2007, just to keep those new light bulbs on for a few hours a day.

This is the scale we're dealing with when we're talking about a billion consumers of any product or service. Thousands or millions of tons of material. Thousands or millions of megawatts. And it keeps going. Think about the raw materials consumed to make those light bulbs, the energy consumed by commuting factory workers, the packaging materials, the ships and trucks used for distribution, and ultimately, the waste that is involved when we have a billion light bulbs. And if we're having trouble delivering a single light bulb to a billion people sustainably, what happens when these billion people want stoves, refrigerators, TVs, computers, cell phones, radios, and cars? What happens when they want street lights, low-cost air travel, hotels, and restaurants? You get the idea.

As engineers, we are already challenged by the environmental impact of products and services today—and the challenge will continue to grow as the world's economy grows. As a result, the scale of our innovation is going to need to meet the scale of the demand for sustainable products and services. We need innovation on many fronts. Using our earlier example, we need the innovation of the compact fluorescent light bulb, which cuts the number of new plants required from 20 to 5; the innovation of better solar and wind-generated power to help us avoid building those plants at all; and the innovation of better product designs using fewer natural resources and more renewable materials.

Eco-Responsible Engineering: An Enormous Opportunity

While much of the focus of this chapter will discuss the negative impact of products and services, we also want to address the potential for positive environmental impact. You'll see that we frequently use the word *sustainability*. Just as *impact* usually focuses on the damage caused by a product or service, *sustainability* forces us to think forward to the desirable end, to ask ourselves what it would take to have a version that could be delivered in massive volume, at reasonable cost, year after year after year. We also like the term because it opens the door to the economic side. If the cost of crude oil rises to $200 per barrel, can we still sustainably sell our product? How about

$300? $400? What effect would a widespread fresh-water shortage have on our service? So, we're looking not only at product sustainability, but also at business sustainability.

An important focus of this section of the book will be the urgency of our global climate situation and the danger posed by steadily increasing releases of carbon dioxide and other greenhouse gas (GHG) compounds into the atmosphere. We, the authors, are convinced by the data that our society needs to make rapid progress in reducing our GHG emissions, including carbon dioxide (often erroneously shortened to "carbon" in the press), methane, water vapor, and so forth.

But we also see that the same behaviors that are creating massive GHG emissions—reliance on petroleum products for transportation, fossil fuels of all kinds for heating and cooling, and coal for electrical generation—are also responsible for a number of our biggest challenges beyond the environment and climate risks. How do we lower our energy dependence on foreign countries, many of which are historically unstable? How do we maintain a stable and economically acceptable cost for energy? How do we meet the energy demands of large, rapidly growing nations such as China and India at a time when oil production appears to be at or near its peak?

So, while we believe that reducing our GHG emissions is imperative, we also see that successfully doing so could provide huge benefits in other areas.

As you read this section you'll notice that we've broadened the environmental discussion to include more than GHG emissions and their impact on Earth's climate. It is a simple fact that the current model of production, consumption, and waste disposal is not sustainable. We are running out of the key resources that fuel our industrial processes: fossil fuels, clean water, and core elements and materials that we use to make our increasingly complex products. Simply reducing GHG emissions won't be enough. We also need to make progress on our other sustainability challenges.

Engineering—possibly more than any other profession—has the power to change the way we interact with our world. True, engineers created the industrial systems that today are pumping billions of metric tons of pollutants into the air and water. But engineers are also capable of designing a new breed of products that reduce harmful emissions, reuse waste materials, and recycle resources. It is within the engineer's power to envision—and create—a new generation of buildings, vehicles, machines, devices, and services that deliver the functionality people want without destroying the ecosystem or depleting scarce resources.

If you're passionate about changing our cultural course and making a tangible, lasting impact, you're in the right place at the right time.

Core Challenges of Eco-Engineering

Eco responsibility remains difficult and uncharted territory for most engineers today, despite the unsustainable nature of today's products and services. Five particular challenges stand out.

- The number of possible environmental impacts is large, and each one can, in and of itself, be difficult to calculate.

- Key impacts of your product may lie outside your company. For example, there may be a large fresh-water impact at one of your suppliers as they make your product, or there may be significant GHG emissions at your customer site as they use your product.

- Most attempts to reduce impacts in one area result in impacts somewhere else. Using wind power is better than burning coal from a GHG point of view, but it involves the manufacture of wind turbines and visual impact to the natural landscape.

- Tradeoffs often involve things that appear, at the surface, to have little to do with each other. For example, what's the cost of cutting down and processing trees for paper bags, versus the short- and long-term waste issues of plastic bags? Many eco tradeoffs are similar, requiring us to make "apples versus oranges" comparisons.

- Engineers think they know how their products will be used, but customers use products how they want to, and transformative products often change users' behavior. Factoring this into product design can be tricky. Could Henry Ford have foreseen the scale of the behavioral change that resulted from widespread availability of affordable automobiles?

Moreover, very little formal training is focused on eco-engineering, despite the avalanche of press about "green" products and eco-friendly design. It is not part of the core curriculum in most engineering schools; it is not a common topic for on-the-job training courses; it's even hard to find a Webinar about eco-responsible engineering.

As a result, it is often difficult for engineers to get started. The next few chapters will help you understand how to work through these challenges and come up with an approach that works for your situation. And there are 7 billion reasons why we really need to get this right.

4

Beyond the Black Cloud:
Looking at Lifecycles

When you're behind a bus that's belching black clouds of exhaust, you can't help but think about the consequences of environmentally insensitive design. It's easy to be disgusted by what the bus is doing to air quality. It's also easy to miss the fact that the billowing fumes represent only a tiny fraction of the environmental impact of the bus. It's harder still to remember that the bus may also have major environmental benefits, despite its obvious eco shortcomings.

Let's start with the total impact. Think about the individual parts that comprise the bus. Thousands of them, globally sourced from a hundred different manufacturers and suppliers around the world, were brought together to a central location for assembly.

Think of the energy expended and the waste created in the process of obtaining the basic materials and manufacturing each of those parts. Consider the greenhouse gas (GHG) emissions of the cargo planes and freighters and delivery trucks that brought the parts through the supply chain and ultimately to the assembly plant.

Think of the fossil fuels burned by workers commuting to and from that final assembly plant—and all of the other assembly plants where the subcomponents were built.

Consider the fact that some of those parts may have used toxic and carcinogenic substances either in the manufacturing process or within the parts themselves, and that those substances will eventually be heading to our landfills. Consider the waste materials left over after the bus has served its useful life: the massive tires with their polyester belts and steel cords, the petrochemical-based seat upholstery, the batteries with their lead oxide plates and rich assortments of acids (all of them with the potential to leach toxins into our soil and water).

And don't forget that the emissions and contaminants and waste of the bus are only one form of its environmental impact—there's also the noise pollution, the damage done to roadways, the visual pollution of the ads and graffiti on the sides of the bus, and so on.

The bus example helps to illustrate two of our engineering challenges. First, the environmental impact of any one product or service is multifaceted and dependent on many factors, and when you're designing something to be "eco-friendly," you need to take all of those factors into consideration. Second, much of the impact of a product may lie outside any single company. If you're the owner of the bus, or even the company that did the final assembly, understanding the manufacturing process or waste issues is challenging at best.

We have no agenda against buses. In fact, they provide a great example of how something that has its own set of environmental impacts can also have a very positive effect: getting cars off the road. If we need to get 40 people from one place to another, a bus is far more environmentally friendly than 40 cars, even after the full accounting of the aforementioned impacts. This illustrates another of our challenges: Even eco-friendly programs, such as mass transit in this case, have an impact that can't be ignored. When justifying a decision based on environmental impact, it is important to consider the full lifecycles of the proposed solution and the alternatives.

This point was driven home by syndicated columnist George Will, who argued in a 2007 opinion column that a Hummer is more environmentally efficient than a Toyota Prius hybrid.[1] He even went so far as to say that "perhaps it is environmentally responsible to buy [a Hummer] and squash a Prius with it." It's a claim that sounds patently ridiculous—until you take a closer look at the concept of "lifecycle analysis."

Yes, the Prius is fuel-efficient by today's standards, but when you factor in the environmental costs of mining and smelting the zinc required for the battery-powered second motor, the production processes for turning the zinc into the component that goes to a battery factory in Japan, and the expected life of a Prius (109,000 miles) versus the expected life of a Hummer (207,000 miles), the environmental benefit of the Prius becomes fuzzier.*

Although some of the numbers behind Will's analysis are suspect—and the Sierra Club's attitudinal advice columnist "Mr. Green" has issued a detailed

---

* There is also a subtler argument that even if today's hybrid lifecycle isn't as favorable as we might expect, future ones are apt to be better. Why? Because the market for hybrids today fuels the R&D dollars for future engineering of what, at a fundamental level, is an innovation in engine efficiency.

rebuttal[2]–the point is well taken: The environmental impact of any product must be measured over its full lifecycle, not just at a moment in time in its useful life.

As we noted earlier, trying to compare the relative impacts of two products or approaches can often be difficult, especially when the impacts aren't directly comparable. For example:

- **Cloth versus disposable diapers:** It's easy to point to the 20-year degradation time frame of plastic diapers and argue for cloth, but when you factor in the environmental cost of producing, distributing, washing and drying, using pick-up and delivery services, and ultimately disposing of cloth diapers, the debate is much less clear. What value do we place on long-term waste versus short-term energy and fresh-water usage?

- **Paper versus plastic:** Consumers have now been conditioned to have a knee-jerk reaction against all things plastic, but when you begin to consider the energy that's used to create a paper bag–cutting the timber, manufacturing the paper, processing the various glues and resins needed to transform the paper into a bag, distributing it to stores, and so forth–plastic may not look so bad.

Sometimes the best answer comes from simply looking at the problem differently. Rather than analyze all of the variables involved with the paper versus plastic discussion, some stores are now promoting a whole new option: asking shoppers to bring a couple of recyclable bags with them to the store. Or consider what Netflix did for the traditional video rental model. Rather than try to find new ways to make the old model more efficient, the company radically changed the whole distribution system and cut the carbon footprint of video rental by orders of magnitude (no more deliveries of new releases to thousands of video stores, no more back-and-forth trips by consumers to rent and return discs).

Because of all of these challenges, we need a framework for thinking through the impacts and tradeoffs of more responsible design. Since we have neither the time nor the data to measure and model everything, this framework must rely on strategies for estimation, prioritization, and focused measuring and modeling within constrained situations. But before we dive in, it's important that we take a step back and ask ourselves what our true goal is here, and make sure that we're not missing the proverbial (and eco-friendly!) forest for the trees.

The "Cradle to Cradle" Vision

So far we've talked about the scale of our environmental challenge and the complexities involved in getting our arms around the problem. But now we need to ask ourselves an important question:

What is the ultimate objective? From an environmental perspective, what are we trying to accomplish?

One common answer: The goal is to minimize the environmental impact caused by this product or service. Fine. *But then we've implicitly said that some amount of damage is acceptable.* There's a presumption that everything that is produced will inevitably create some quantity of waste, and the best we can do for Mother Nature is to reduce the volume of that waste and take greater care in disposing of it.

But the potential exists to get beyond an approach that simply minimizes damage. In their remarkable book *Cradle to Cradle*[3] William McDonough and Michael Braungart have laid out a compelling vision that does just that.

The book begins with the premise that the processes we have today for designing and building things are linear and one-way—a cradle-to-grave approach. Our companies make what consumers want and get it to them as quickly as possible, while waste is assumed, and disposal of waste is simply part of the process.

McDonough and Braungart contend that making this model more efficient—focusing on reducing, reusing, and recycling—is not sufficient, that it perpetuates a deeply flawed model and merely delays the inevitable: immense damage to the environment and depletion of vital natural resources. In other words, being less bad is not good enough.

> Eco-efficiency is an outwardly admirable, even noble, concept, but it is not a strategy for success over the long term, because it does not reach deep enough. It works within the same system that caused the problem in the first place, merely slowing it down with moral prescriptions and punitive measures. It presents little more than an illusion of change. Relying on eco-efficiency to save the environment will in fact achieve the opposite; it will let industry finish off everything, quietly, persistently, and completely.

The book challenges engineers to focus not on eco-efficiency but on "eco-effectiveness," or designing products in a way that actually replenishes and nourishes the environment rather than simply using up natural resources. One example of eco-effective design would be a new kind of roofing. As opposed to the traditional layer of asphalt shingles or wood shakes, an eco-effective roof would be a light layer of soil covered with plants. Its "growing grid" would maintain a stable temperature, provide cooling in hot weather and insulation in cold weather, and last longer because it would shield underlying building materials from the sun. In addition, it would produce and sequester carbon; it would also absorb storm water.

The vision articulated in *Cradle to Cradle* is compelling and inspiring. But is it practical? Do we, as engineers, have the creativity, the resolve, and the resources required to make the leap from the cradle-to-grave approach to this new model?

Some of you may be thinking that it would take a catastrophic event—the submersion of Manhattan or Silicon Valley due to polar ice-cap melting, for example—to get people to make such fundamental changes. Others believe profound cultural change requires a unity of purpose that's no longer possible given the diversity of our society. Still others are thinking it's already too late and our collective doom is inevitable.

We don't think the cradle-to-cradle vision is an all-or-nothing proposition. Our advice is to use it as a guiding principle. Look for opportunities to fundamentally rethink product and service design in a way that conforms to the vision of eco-effectiveness. We believe that as you attempt to put this vision into practice you will find more and more ways to bring it to reality. However, we also have to be realistic and understand that we sometimes don't have the time, technology, or know-how to apply this vision to every situation. In these cases, we need to make sure that we still practice eco responsibility, reducing impacts as much as we can along the way.

The key is to allow yourself to be more creative in how you approach product design. As we've said, sometimes the most eco-effective alternative comes to light only when you look at an engineering problem from a completely different angle. Consider the example of Interface Carpet. Several years ago, engineers at the company were trying to find an eco-friendly adhesive for a popular new line of modular carpet tiles and carpet squares. For almost two years they experimented with many different compounds, looking for one that provided the right level of adhesion, protected the carpet against molds, didn't wreck the underlying floorboards, and didn't use toxic chemicals. In the end, their solution was not to use an adhesive at all. Instead, they used gravity and rubber, a renewable natural resource that met all the design criteria with minimal environmental impact.

Another example from our own experience is the development of "chip multithreading" (CMT) processors. The obvious way to get higher performance from a processor is to make the "clock" go faster, and for the past few decades that's been the standard approach. Sun engineers, however, found a way to improve overall performance by enabling a single processor to execute more "threads" or tasks in a given period of time. That way, the processing power per thread could actually be reduced—resulting in lower total power consumption—while overall performance increased substantially.

So, we'll conclude this chapter with this thought: Eco-effective engineering is not simply "visionary," and it is not outside the scope of pragmatic engineers. The vision articulated in *Cradle to Cradle* can become reality when there's an intersection of will, creativity, and the right opportunity. But can we count on a true cradle-to-cradle breakthrough on every project? Can we hit a home run on every trip to the plate? Unfortunately, no. So, for the remainder of this part of the book we're going to focus on what to do with the impacts that you're left with after you've tried your best to reach the ideal.

5

A Pragmatic Approach to Lifecycle Analysis

Formal lifecycle analysis is not new; in fact, lifecycle analysis tools and techniques have been around in various forms for decades. What is new is an urgent need to improve the tools and expand the use of lifecycle analysis to a broader spectrum of products and services.

We're going to use a pragmatic approach to lifecycle analysis that keeps the focus on the main goals: understanding the overall impact and making improvements. The truth is that you don't always need to measure everything; you don't always need precise data; you don't always need complete information. You just need to know what to measure, when, and how—and where to place your priorities.

To get started we'll need a model of the product/service lifecycle that we can use to organize our work. So, let's take a closer look at the phases of a typical lifecycle and the key considerations at each phase.

A Basic Lifecycle Model

Every product is different; every lifecycle has unique time frames and characteristics. As a result, many different lifecycle models have been produced over time. For this book, we use a basic three-stage model. We prefer this model because it is straightforward and matches most people's personal experience with the lifecycle stages of common products. The three stages of our model are

- "Make," which covers everything that happens before a product is actually put into operation—including the materials and chemicals that are used to create it, the processes involved in assembling and manufacturing it, the packaging that encases it, and the supply chain that distributes it

- "Use," which includes the power the product consumes as it is operated, the greenhouse gas (GHG) and other emissions it creates, the water it uses, and the noise, light, and heat it generates during operation

- "Renew," which covers everything that happens after the product is used, including the demanufacture or disassembly of the product, reuse of key components, recycling, and take-back

At each stage of the lifecycle we focus on three primary aspects of the environmental impact of a product or service:

- **Energy and emissions,** including the calculation of energy and power, finding the cleanest source of energy for your product, using energy efficiently, calculating GHG emissions and CO_2 conversion, and so on

- **Chemicals, materials, and waste,** including the legal and business considerations of hazardous and toxic substances, packaging and doc- umentation, waste disposal, recycling, take-back, and process-related GHG emissions

- **Water and other natural resources** that are embodied in the product or service, including social and business considerations of using scarce or nonrenewable materials, calculating the water footprint, and so forth

Additional Lifecycle Considerations

Our three-phase model is intentionally simplistic. So, before we discuss each aspect of the lifecycle in more detail, we'd like to offer a few notes about topics that our model doesn't directly address, including supply chains, consumables, hidden impacts, services, and the effects of design and prototypes.

Supply Chains

In this day and age, very few products are manufactured and distributed by one company. Most are produced and marketed through complex ecosystems of geographically dispersed companies. To engineers this means that when you're designing products and services, every participant in the supply chain matters: your company's employees, your company's suppliers, those suppliers' subcontractors—everyone involved in assembling, manufacturing, distributing, and ultimately disposing of your product.

The boundaries of accountability are expanding, and this accountability is so new that most industries and companies haven't yet developed the tools, standards, or certifications needed to fully address it. However, many companies that are committed to global citizenship are beginning to formalize supply chain environmental and social responsibility programs, supplier code-of-conduct contracts, supplier site audits, and more stringent product requirements and specifications.

This presents an opportunity for engineers to set an example, which can directly impact the performance of your suppliers. For instance, engineers at Tesla Motors (the Silicon Valley company that's producing high-performance, 100% electric cars that we talked about in Chapter 2) are conscious of everything environmental—"from the design specs of the automotive systems and software they're working on to the temperature of the building, whether the lights are on, [and] whether you drink water out of bottles or out of the tap," says Craig Carlson, Tesla's director of software engineering. "Eventually, this translates to an ability to migrate suppliers to a more stringent code of conduct that follows our example."

We're also beginning to see more collaboration within various vertical industries centered on lifecycle supply chain issues. For example, the electronics industry has banded together to help improve socially and environmentally responsible performance across manufacturing and supply chains. Two major working groups in that field are collaborating: the Global e-Sustainability Initiative (GeSI) and the Electronic Industry Code of Conduct (EICC).[1] These types of collaborations are important in larger industries where suppliers often sell to more than one of the major manufacturers. Without alignment, such suppliers may be faced with trying to meet a set of different—and potentially contradictory—environmental goals and reporting standards.

Finally, from a practical point of view, how do you get sustainability information about suppliers? The place to start is with the group that manages suppliers within your company. They may have data and mechanisms already in place. Also, it is always important to make sure they are aware of discussions you are having directly with suppliers, as the supply management team owns the contractual obligations between your company and the supplier.

Beyond that, there are two ways to get data, and you will often find yourself resorting to a mix of the two. First, you can simply request the data from your suppliers. The second approach is to use externally generated models, which may not provide information about a specific supplier, but can characterize a general process or industry. For example, you may use the Economic Input/Output (EIO) models that we discuss later in this chapter. Another example of this approach is paper; well-established models exist for calculating the impact of paper lifecycles without consulting specific manufacturers.[2]

Gathering data one supplier at a time can be a complex and time-consuming task, but if you do it correctly (and if suppliers are cooperative), you can get very accurate results. External models are often much simpler and quicker, but may miss important, supplier-specific details.

"Mini Lifecycles" of Consumables

Many products don't fit the simple lifecycle model outlined in the preceding section because they contain elements or subsystems that have their own separate lifecycles, and it's important to factor these in when performing a lifecycle environmental impact analysis. Examples include

- Batteries, toner cartridges, ink refills, filters, and other consumables.

- Transport packaging, bulk shipping packaging, retail display packaging, consumer packaging, packaging that holds other packages together, and so on.

- Manuals, technical guides, troubleshooting references, warranty cards, and so forth.

The approach for dealing with these is pretty straightforward: Do a lifecycle analysis for each one and add them up. In the case of the printer, you can

1. Work through the impact of the printer unit itself.

2. Add in the lifecycle of an ink cartridge multiplied by the typical number of cartridges used in the lifetime of the printer.

3. Add in the impact of the packaging and documentation.

4. Do the same thing for the paper usage (try the Environmental Defense Fund's paper calculator at www.papercalculator.org).

From a pragmatic point of view, it is worth coming up with a rough estimate of each component before spending too much time on the specific measurement of any given piece. If you find that the energy use of one component is a thousand times greater than any other, from an energy perspective it's probably a good idea to focus on the part with the greatest energy usage and ignore the others until they become more significant.

Hidden Impacts

Some products or services have environmental impacts above and beyond the readily observable direct impacts. Here are just a couple of examples.

- If your product runs on electricity or is often used in an air-conditioned environment (such as a PC in an office) it will usually generate some heat as a result of inefficiencies in the electrical components. Therefore, you need to account for the environmental impact of the air conditioning required to compensate for its operation.

- Many products require a professional installer or regular visits from a service technician. The environmental impact of the technician's travel to and from the site should be accounted for, as well as spare parts that are needed.

- Some products have special disposal requirements. The mercury in compact fluorescent light bulbs requires that they be recycled at specially equipped facilities that can deal with the mercury.

Generally, these situations aren't hard to calculate; you just need to spend some time and make sure you think of these "hidden" impacts for your specific situation. And again, from a pragmatic view, you may be able to determine that some of these impacts are small enough to be ignored. If a product travels 10,000 miles from China after it's made, and then has to make one last trip of an average of 2 miles to be recycled, you can probably safely ignore the impact of the recycling trip for now.

Services

The lifecycle model applies to services as well as products. In fact, sometimes services can have a major impact when they are done on a large scale (remember what happens when we multiply anything by 1 billion). Here are a few types of services, and tips on how to approach them.

Traditional services, such as consulting engagements or technical support services, often involve sending trained specialists or replacement parts to customers. As a result, these services are often dominated by the travel and shipping involved. Services can also involve call centers, where the impact is from the office space, commuting, and equipment required to support the call center operation. In any of these cases, the impacts are fairly straightforward to calculate based on travel or shipment calculators, or from energy bills.

Web-based and online services, such as search engines, online shopping sites, or massively multiplayer online role-playing games (MMORPGs) such as World of Warcraft or Lineage 2, often involve massive compute power from very large data centers. In one notable example, the avatars or virtual characters in the online 3D world called Second Life were actually determined to have a carbon footprint almost as big as that of a typical real person in Brazil!*

Some products have both a product and a service element. One example is a Webkinz stuffed toy, which comes with a special code on its label that allows access to the "Webkinz World," a Web site where kids can interact with a virtual version of their stuffed pet (see Figure 5–1). The maker, Ganz, has sold more than 2 million units to retailers and claims more than 1 million registered users of the Webkinz Web site. To assess the full lifecycle impact, you have to add together the impact from the manufacture and delivery of the physical good along with a pro rata share of the online infrastructure.

Another example is TiVo, a digital TV recording system that includes a physical box that attaches to your TV, as well as a back-end online service that is responsible for feeding the box TV schedules and software updates.

The key point is to understand the environmental impact over the entire lifecycle—both the product and the service components must be accounted for.

Design and Prototypes

As engineers, we know the true lifecycle of a product begins well before the manufacturing stage. Prior to that, we're writing specs, doing design, building prototypes, testing, fixing bugs, and testing again. In a rigorous lifecycle

---

* Source: Author Nicholas Carr. Carr calculated that, on average, there are about 12,500 active avatars on Second Life at any given time of the day, and that 4,000 servers and cooling systems are used to support the world, combined with the 12,500 PCs that are used to control the avatars. That amounts to 1,752 kWh of electricity used by each avatar over the course of one year, and the average citizen of Brazil consumes 1,884 kWh, writes Carr (read Carr's blog at www.roughtype.com/).

FIGURE 5-1   The Webkinz Web Site

analysis, you'd need to fully account for all of these activities. However, for most real-world products, the premanufacturing impact is very small compared to the impact of full production and usage.

Of course, in some situations this may not be the case, and you will need to include design and prototyping in your model and accounting. For example, only a few copies of the space shuttle have been built, so the effort to build prototypes of the pieces may be nontrivial compared to the production runs. Similarly, the number of cars smashed in crash tests may be worth counting for cars with smaller sales volumes.

How do you decide whether you need to count the design and prototype impacts? As in earlier cases, do a quick analysis and see whether you can gauge the relative sizes of that impact versus the production and usage impact. If it's less than 1%, you can ignore it unless you're doing a very detailed model. If it's more than 1% of your overall impact, it is worth at least tracking and paying attention to possible ways to reduce it.

Embodied Energy and Embodied Carbon

As you do more work in lifecycle analysis you will come across the concept of **embodied energy,** and related concepts such as **embodied carbon** (which really is inaccurate shorthand for embodied carbon dioxide equivalents [CO_2e], which we will discuss in a later chapter).

Embodied energy is a measurement (or, in reality, a modeled estimate) that represents the total energy required to manufacture a product and get it to the customer. This concept provides a useful way to understand the impact of the parts of the product lifecycle that are usually not visible to the consumer. If someone tells you the embodied energy of a product, you can get a sense of the scale of energy impact of making that product without having to know the details of how it was made and shipped.

The concept of embodied carbon is similar to that of embodied energy—it's an aggregate of the direct and indirect carbon emitted for the entire production process. The methodologies for calculating embodied carbon are also similar to those used for embodied energy, though there tend to be more areas of ambiguity in how things are accounted.

However, standards are emerging for measuring embodied carbon within individual companies and across the supply chain. For example, the World Resources Institute's GHG Protocol Corporate Standard[3] provides standards and guidance for companies and other organizations preparing a GHG emissions inventory. It covers the accounting and reporting of the six GHGs covered by the Kyoto Protocol, and was designed to help companies prepare a GHG inventory that represents a true and fair account of their emissions while simplifying and reducing the costs of compiling a GHG inventory. We make use of this standard in our free, online GHG tool, OpenEco.org, which we will discuss in detail in Chapter 11.

In addition, the **Carbon Trust,** created by the U.K. government, is developing a draft Publicly Available Specification (PAS) standard by which the carbon content of all products and services can be measured. The final PAS standard will be a useful tool for engineers to identify and analyze GHG emissions associated with the products and services their companies provide. It draws on lifecycle assessment (LCA) techniques and can help identify and quantify the environmental loads involved, the energy and raw materials used, and the emissions and wastes consequently released.

Lifecycle Assessment Tools

Since lifecycle analysis has recently become such a hot topic, it's not surprising that there are tools you can work with today—or that significant work is underway to create new tools. While you won't be able to rely on any single tool to do all of your analysis work, the important thing is that learning to use the tools can be very helpful and illuminating, even if they don't always give you complete, consistent, and precise results. Our general advice is to start using the tools that are available, come up with some consistent methods for measurement, and present your findings openly and transparently, no matter what they are. Remember to keep track of which tools you used for which estimates. As tools and accounting methods improve, this information will help you understand how accurate your results are.

On one end of the spectrum are tools based on the **Economic Input–Output Life Cycle Assessment (EIO-LCA)** method. This method is based on macroeconomic data, so it's great for getting a very quick overview of the impact of a class of products. It is also unique in that it includes a broad range of impacts, including natural resources, energy, and GHG emissions. In that regard, EIO-LCA can help you understand the relative sizes of various impacts and a top-level estimate for each. It's also a great way to get an estimate when no other method is viable.

On the other hand, the EIO-LCA method is primarily based on macroeconomic data, which makes it very difficult or impossible to examine individual product traits or tradeoffs. For example, if you make a change to your product, the resulting change in impact won't show up in an EIO-LCA model for years (if ever).

The mathematics of EIO-LCA analysis is straightforward, but the model requires that someone has done a detailed analysis of the industry and product classes in question. This renders the tool impractical in some instances, because specific industry data may not exist or may be incomplete or inaccurate. However, when the right data exists, an EIO-LCA model can provide very complete results. Want to understand the impact of the trucks that delivered parts to your suppliers? An EIO-LCA model won't tell you how many trucks were involved, but it can give a good estimate of the impact.

For an excellent tutorial on the economic input-output analysis method, we suggest you visit the Eiolca.net Web site sponsored by Carnegie Mellon's Green Design Institute at www.eiolca.net/tutorial-j/tut_1.html. This tutorial provides step-by-step instructions on how to enter data into the EIO-LCA model and the process behind choosing the data to enter. The tutorial also describes how to read and interpret results. We also recommend the technical white paper by Chris T. Hendrickson, Lester B. Lave, and H. Scott Matthews, "Environmental Life Cycle Assessment of Goods and Services: An Input-Output Approach."[4]

The main alternative to EIO-LCA for measuring environmental impacts is the **process-sum** approach. It's based on process and facility-level data that quantifies material inputs (consumables), material outputs (products), and emissions over the three lifecycle phases. In other words, you calculate the energy expenditure for each discrete activity involved in making, using, and renewing the product, and you add them all up. For a given product there may be hundreds or thousands of separate calculations, but the process-sum analysis can provide a fairly accurate assessment of embodied energy for the processes or products being analyzed.

The EIO-LCA and process-sum methods actually complement each other quite well. EIO-LCA is good at getting an estimate of the total impact, but is weak on the details. Process-sum is strong on the details, but it requires a big effort to build a complete picture. As a result, we highly recommend using a mix of both methods, with the specific tool being chosen depending on what question you're trying to answer.

Because these methods are so complementary, a new method that formally combines the two has emerged. The **separative hybrid** approach supplements process data with I/O analysis and uses estimates where hard data is not available. This approach is fairly new, but it has shown promise in more accurately gauging the embodied energy for consumer products such as desktop computers or refrigerators.[†]

In many cases, you can make simplifying assumptions about certain elements of the model to get a meaningful result with less effort. For example, when you're considering the energy expended during shipping you could just use the whole weight of the product shipped *n* number of times, as opposed to summing all of the individual components at various times during the process and trying to track exact values at multiple times. Or you may be able to extrapolate good data from previous studies. Lifecycle studies of PCs, for example, may contain data that's helpful in understanding the eco impact of everything from cell phones to servers to DVRs to the electronics in cars. Leveraging other people's work as a starting point can help you take a huge step forward in analyzing your specific project.

[†] For details about the economic I/O and separative hybrid techniques, and a case study using the methods to calculate embodied energy for a desktop computer, see the white paper "Energy Intensity of Computer Manufacturing: Hybrid Assessment Combining Process and Economic Input-Output Methods," by Eric Williams, published in *Environmental Science & Technology* 38(22): 2004.

Finally, lifecycle modeling often involves straightforward calculations that can be done simply if you just have the right constants to calculate with. For example, burning gasoline in a standard internal combustion engine creates 19.9 pounds of GHG emissions (measured in CO_2 equivalents) per gallon of gas used. This convenient fact makes it easy to know the GHG emissions of operating any gas-powered vehicle; you don't need to know the weight, the engine size, the miles per gallon—just knowing the fuel involved is enough!

Most other tools provide insight into particular products or product elements, and are sometimes limited to a smaller number of impacts (e.g., GHG emissions only). The paper calculator that we discussed earlier is a prime example, and other similar calculators are becoming available.

One interesting example, created by the Flemish Waste Agency, is a set of cards that makes it easy to compare the lifecycle environmental impacts of commonly used materials and processes.[5] The "Ecolizer Designwijzer" cards contain several hundred eco-indicators quantifying the environmental impact of the three lifecycle phases of materials and processes, including ferrous metals, nonferrous metals, polymers, wood, paper and packaging, building materials, chemicals, energy, transportation, as well as buildings, land use, lighting, and services (see Figure 5–2).

FIGURE 5–2   The Ecolizer Designwijzer Tool

The method used to create the indices looks at the damage caused in terms of resource depletion, land use, climate change, ionizing radiation, acidification/eutrophication, and toxicity. This tool is mainly intended for quickly estimating relative comparisons of products and components. The guide is written in Dutch and currently is distributed only through workshops, but it offers an example of how governments everywhere could make getting eco-design know-how into people's hands in forms they can use.

Other examples of recently developed lifecycle analysis tools include the following.

- **TEAM:** This is an LCA tool from Ecobilan (a.k.a. Ecobalance), which allows the user to build and use a large database and to model any system representing the operations associated with products, processes, and activities. TEAM can be used to calculate the associated lifecycle inventories and potential environmental impacts according to the ISO 14040 series of standards (for more information see www.ecobilan.com/uk_team.php).

- **LCA calculator:** This tool, from Industrial Design Consultancy, helps users assess the environmental impact of a product by calculating its energy input and carbon output (for more information see http://lcacalculator.com/).

- **GaBi 4:** Available from PE International, this provides sustainability data administration and evaluation at the organization, facility, process, or product lifecycle level.[6] Technically GaBi 4 is not a tool, but a set of tables of useful numbers (for more information see www.gabi-software.com/english/gabi/gabi-4/).

Starting a Top-Level Assessment

Some engineers like to immerse themselves in understanding the formal process before engaging in a problem. Others (especially certain software developers) like to dive right in and learn by doing, adding process as they get further along. Those who use the latter approach may even start from the beginning more than once, using what they learned on the last try to tune their approach.

This section is for engineers who like to jump right in. We're going to walk through the process of starting a lifecycle assessment based on what's in your head, and we suspect you'll realize that you know more about the sustainability of your product than you might have thought.

The first thing to do is to open a text file (or just pull out a pencil and paper if that's more natural for you) and write down the three major lifecycle phases: make, use, and renew. For each phase, simply write down where potential impacts may lie for each major category of impact: energy usage, chemicals, emissions and waste, and water and natural resources. At this stage, this should just be a brainstorming exercise—don't go crazy with numbers, just try to identify any area that may have a significant impact.

Next, review the special cases we covered in this section: Are mini life-cycles involved? Are there any services? Are there any indirect or hidden impacts?

Now identify the scale involved. At this point, you probably don't know the size of the impacts, but you should be able to describe the size of the process that is causing the impacts. Here are some examples of what you might write down.

- "Water and natural resources to create 100,000 copies of the documentation"

- "Energy and resulting emissions to ship 500 units from our factory in Mexico to customers in Europe"

- "Electricity to run 50,000 units for one year"

- "Fraction of the product that is easily recyclable aluminum or plastic"

- "Impact of making and disposing of 1.2 million batteries"

This level of information provides you with a basic model for diving in and understanding where your impacts are and what to do about them. Each of these can be worked through and converted to more specific numbers (e.g., actually calculating the water involved to make the paper in the first bullet). We'll discuss how to do these calculations in more detail in later chapters.

Next, for each item you listed, write down some additional notes where applicable. For example:

- **Estimate the cost of the activity.** For instance, estimate the shipping cost, electricity cost, cost of materials, and so on. This will provide a basis for finding the potential cost savings associated with eco improvements.

- **Look for product benefits.** Are any of the items you listed valuable to your customers? Do they view them negatively? This will highlight potential areas for feature improvement.

- **Consider legal implications.** Obviously, you need to do a detailed analysis of applicable laws, but even at this point it's worth noting the areas where you suspect or already know there are important laws or standards.

Before we go too much further, it's useful at this point to check your work against some other models. In some cases, you can find published analyses of products that are similar enough that you can do a sanity check, and some quick work with the EIO-LCA models might also give you a sense of whether you've missed anything big. While there may not be a perfect EIO-LCA category for your product, you can look at a similar product, or apply the EIO-LCA model to subcomponents of your product or service that you identified earlier.

Deciding When to Stop Assessing

This is one of the tough questions of lifecycle assessment. You need to be clear what your goals are, and then think through how good the data needs to be to meet your goals. At a minimum, you need to

- Be able to communicate where your biggest impacts lie and how big they are

- Have sufficient data to meet any legal claims you'll be required to make

- Be able to use your model to understand the size of the impacts as a result of design changes in your product or service

For some situations getting within a factor of two for the overall impact in the key areas may be good enough. For other projects you may find that you need to be accurate to within one or two percentage points. And in some areas, such as the level of potentially dangerous chemicals, you may need to be even more precise. Note that the answer doesn't have to be the same for each part of your project. The precision and accuracy needed to meet a legal requirement will not be necessary in other areas.

Finally, it's important that you keep track of how good your data and model are in each area. For example, you may have decided that a factor of two was sufficiently close for one purpose, but then find that your marketing department wants to make some strong public claims based on your numbers. It's critical that you keep yourself honest on what the numbers really mean, and make sure they aren't being misused.

At this point, you've created an assessment model and you can iterate on your model until you're comfortable that you've captured the major impacts to a level of detail required for your project. So, now we'll shift gears from the model itself to determining where to focus first.

6

Setting Priorities, Requirements, and Goals

With a rudimentary lifecycle model such as the one we constructed in the preceding chapter, you can see where your product or service will have impacts. The next question is what to do about them. In general, you'll find that there are some areas where you *must* make improvements, some areas where it's advantageous to make improvements, some areas where it's easy to make improvements, and some areas where it won't be worth your time and effort.

Once you determine your priorities, then you can establish the requirements and goals for those areas. In setting priorities, we have found that it is useful to look at your product or service from four different points of view.

- What are the legal requirements?

- What are the business requirements or opportunities presented by your target market?

- What are the largest impacts?

- Which things would be easy to fix?

Each of these perspectives will provide a different insight regarding where to focus, and will provide a framework for establishing requirements and setting goals. Let's look at each of them individually.

Knowing the Law

If you're on the wrong side of the law, nothing else will matter. It's critical that you understand all the regulations, restrictions, and government policies that will apply to your product or service so that you can achieve—and sustain—legal compliance. As they say, ignorance of the law is not an excuse.

Typically, laws fall into the following categories:

- Banned or restricted chemicals, toxins, and hazardous substances

- Directives on packaging and packaging waste

- Product take-back, recycling, and waste disposal regulations (such as the European Union Waste Electrical and Electronic Equipment [EU WEEE] Directive)

- Emissions limits, including electromagnetic radiation, noise, other emissions, or exhaust

Laws in these areas vary by country, state or province, and even municipality. In short, you need to understand the legal landscape in any market where you will potentially sell your product or service.

While this sounds daunting, there are lots of resources to help you. Industry trade groups and consultants exist for all major industries and can help you understand the overall landscape as well as the details of specific laws. Most large companies also have internal compliance groups, and often have established a common requirements list that is known to be compliant with all target markets. So, don't try to figure these issues out on your own; it's a very complex topic and there's no reason to take the risk of missing a key law or misinterpreting the applicability or requirements of some piece of legislation.

One important question to answer at this point is whether you are going to make one product that you will sell worldwide, or whether there will be separate versions that comply with specific laws on a regional basis. For example, today many electronics companies make global products, but the differences in regulations cause car manufacturers to make and sell different, region-dependent models. This is a really big decision that touches many parts of the organization. If your company doesn't already have a policy on this, you need to spend some time and make sure you've devised a plan that will work for your company.

One challenge with environmental compliance today is that many new laws are under development. Since products are often in development for years, you will need to make sure you're up to speed on potential laws that may be in effect when your product eventually ships. In some cases, you or your company will need to get involved in the legislative process, educating government officials on the nature of your products or market. You may find a need to lobby for exceptions, help with the definition of product classes (which products are covered by a law and which aren't), and influence the levels, quantities, or thresholds specified in the legislation. Again, you may do this directly, or you may find that others in your industry share the same issues, in which case it may be more efficient and effective to engage the government through an industry group.

Finally, if you are developing something in an emerging product space, you need to pay extra attention to which regulations are or aren't applicable to you. Is a digital video recorder (DVR, e.g., a TiVo) a computer or a piece of consumer electronics, like a DVD player? Are the new "cross-over" vehicles cars or SUVs? Are thin desktop systems the same as PCs or are they just smart monitors? If you have any doubt, engage the applicable government agencies and get a formal ruling. Nothing would be worse than finding out your awesome new product can't be sold in some country or region because the government there viewed your product as being in a different class than you did, so as a result you used the wrong set of design requirements.

At Sun, we started paying much more attention to the legal aspects of environmental responsibility beginning in about 2000, when it became clear that local, regional, and statewide hazardous substance laws were gaining momentum. We saw that the EU RoHS (Restriction of Hazardous Substances) Directive initiated in Europe was likely to be implemented globally in some form, and we understood that RoHS would continue to evolve, expanding to encompass more and more of the REACH model (Registration, Evaluation, Authorization, and Restriction of Chemical Substances). At different points in time, we've looked at whether we needed regional products, but have continued to stay the course and produce products that meet the sum of laws around the world so that they can be sold anywhere in the world. This is important to many of our large customers that have IT operations around the world and want to be able to standardize on specific product models and use them everywhere they operate.

At the same time, we began paying more attention to regional and local recycling regulations, such as the WEEE initiative launched in Europe. We also began focusing more resources on complying with Energy Star guidelines and EU Energy-using Products (EuP) directives. Finally, we've been spending a lot more time in Washington, Brussels, Beijing, and elsewhere lobbying for consistency in regulations and standards. Any differences between countries, regions, states, and so forth can significantly raise the cost of doing business, often with no added benefit to the planet.

Business Requirements and Opportunities

> "It's more obvious each day that extreme efficiency is good
> for the environment and great for business."
> *—Jonathan Schwartz, president and CEO, Sun Microsystems*

Corporations are recognizing that environmental responsibility and fiscal responsibility are not mutually exclusive—that, in fact, they can mesh well. And that's great for engineers who want to improve the eco-effectiveness of a product or process. Of course it also means you have to be up to speed on the economic impact of being ecological.

In some cases, changes to a product or service to address ecological issues or opportunities will be synergistic with business. They will reduce costs for you or your customers, or may add capabilities or features that grow the market appeal of a product. For an engineer, this is fun stuff because it's a double win—you're helping the environment and the company's bottom line at the same time. It is important to recognize these opportunities and maximize the business upside. When you look at each of the processes and impacts of your lifecycle model you should ask yourself: Is there an opportunity for us or our customer to save money? Can we make the product more desirable?

In other cases, changes to a product or service will have negative economic impact. Often this is because costs will have to go up without any upside in revenue. Other times this is because the need to implement an environmental feature will reduce the desirability of a product in the market (does anyone really want to be troubled with taking his light bulbs to a specialized recycler?). When this happens, engineers need to be aware and look to minimize the negative business impact.

Finally, many corporations today have begun to formalize specific environmental objectives. For example, your company may have a public goal to decrease its greenhouse gas (GHG) emissions or water usage by a certain percentage by a specific date. Or the company may have overarching product goals to decrease packaging or increase product energy efficiency. It is important to understand these objectives and factor them into your project plans. Helping the company meet them may give your project some extra resources or support, whereas ignoring them may keep the company from meeting its goals and bring problems to your project.

Businesses are clearly eager to participate in the green movement. But that doesn't mean your particular project will sail through your company's approval and budgeting processes simply because it's eco-friendly. From the company's point of view, it's still about money. You need to be specific about how your project will solve customer problems, contribute revenue, and create competitive advantages.

For engineers, this is an enormous opportunity. By mastering the economics of eco-effective design, you can play a crucial role in improving the environmental impact of products and services throughout the lifecycle—from using more eco-responsible source materials, to creating more eco-efficient manufacturing processes, distribution techniques, and packaging materials, to inventing more cost-effective and environmentally sustainable disposal methods.

Many books and dozens of Web sites are available to lay out the business case for eco-effective products and services. Among the best:

- *Green to Gold: How Smart Companies Use Environmental Strategy to Innovate, Create Value, and Build Competitive Advantage* by Daniel C. Esty and Andrew S. Winston (Wiley)

- Business of Green: A Global Dialog on the Environment, a blog published by the *International Herald Tribune*

- GreezBiz.com, a leading online information resource on how to align environmental responsibility with business success

You don't need an MBA to get a handle on the economic advantages of eco-friendly design. You *do* need to understand and communicate the advantages of your particular project from a business perspective.

Areas of Greatest Impact

Among the various impacts of your product or service, one or two typically stand out as being significantly larger than others. If you burn coal or petroleum, maybe it's CO_2 emissions. Or maybe you use a manufacturing process that requires lots of fresh water or some other key natural resource. Or maybe it's a specific hazardous chemical that you require.

Whatever it is, you need to understand the areas where your product has its largest environmental impact and start working to minimize that impact, no matter how hard that may be. For one thing, it's the right thing to do. But beyond that, if it is a major impact for you, sooner or later someone will start asking questions and it will be important that you have recognized the impact, can talk about it coherently, and are executing a plan for reduction.

Consider an example from our own experience at Sun. A few years ago we saw that electricity costs represented an increasingly large share of our customers' IT costs, and that customers were having trouble getting sufficient electricity for their data centers. With the rapid build-out of Internet-based services and new Web 2.0 applications, the deployment of new infrastructure was outpacing the ability to provide adequate power and cooling of equipment.

A back-of-the-envelope calculation showed that it was by far our biggest environmental impact—many times the energy and emissions required to make our product or run our own operations. Furthermore, we understood that it was a growing issue for our customers, and they were going to expect progress. We began investing—and innovating—more in areas such as energy-efficient chip design, core operating system efficiency, and innovation in data center consolidation techniques and virtualization technologies. Our coming-out party occurred in 2005 with the launch of the "Niagara" processor, a new UltraSPARC architecture that used only 74 watts of power and ran at 1.4 GHz—far faster than previous generations. Industry analysts have credited this innovation with saving customers many millions of dollars in energy costs—and catapulting Sun into new business opportunities.

Quick Wins and Low-Hanging Fruit

Through analysis and measurement, not only will you identify opportunities for making major improvements in terms of environmental impact, but you'll also run across areas where quick fixes can be made with minimal effort or expense. When you find such low-hanging fruit, go ahead and pick it. Quick fixes may not have a huge impact individually, but lots of them add up.

For example, there may be simple things you can do to reduce the weight of your product. The weight has huge ripple effects in terms of environmental impact: Every ounce represents more raw materials used, more energy expended to produce and assemble and distribute your product, more GHG emissions (especially when you factor in multistage shipping), more waste to deal with, and so on. Reducing the weight of your packaging or documentation by a small fraction can provide a nice reduction to the total carbon footprint of your product.

Some of these opportunities will become apparent in your modeling activities, but others will bubble up from your company's employees. They are the ones who know how things really work, sometimes in ways that aren't ideal.

We get a steady stream of emails that start with "Did you know...", and some of our best quick wins have come from such employee observations and suggestions.

In addition to the environmental benefit, we often find that small, quick wins can provide good momentum to the organization, and provide stimulus for tackling some of the tougher problems. Also, focusing on low-hanging-fruit projects with a positive economic return (lower shipping costs, for the weight reduction example we just discussed) is a good vehicle for engaging the finance department and business management.

So, to quickly recap our guidelines for setting priorities, requirements, and goals:

- Make sure your product or service will meet all applicable environmental laws for every country, state, and province where you plan to offer it.

- Place a high priority on environmental features that will cut costs or increase market appeal for your product or service.

- Place a high priority on minimizing the effect of any environmental features that are required but will not have a negative business impact.

- Place a high priority on environmental features that will help your company meet its overall environmental goals.

- Understand the largest impacts of your product or service and make sure you have a plan to decrease the impact steadily over time.

- Be on the lookout for low-hanging fruit—opportunities to make changes that have a positive business impact, no matter how small they are.

7
Energy and Emissions

This chapter provides basic information for understanding energy and calculating emissions, including common sources of energy, calculating energy and power, impacts of various power sources, batteries and standby power, measuring greenhouse gas (GHG) and other emissions, and additional topics such as energy usage and efficiency opportunities in data centers.

"The most available, cheap source of energy is that which we waste."
—*Alexander ("Andy") Karsner, former U.S. Department of Energy Assistant Secretary for Efficiency*

Energy is at the heart of every product or service. Every day more products depend on electricity to operate and more petroleum-based fuels are used to transport products to their customers (and, hopefully someday, to the recycling center). Even purely digital products have an energy component. It has become hard to imagine a useful product or service without energy.

There are three important things to remember about energy.

- Energy is used at every stage of the product or service lifecycle. While the energy consumed during product use is the most visible, it may not be the largest component of energy use in the whole lifecycle.

- All sources of energy have impacts that need to be taken into account, ranging from GHG emissions to security to safety to impacts on natural spaces. While we often refer to certain types of power as "green," there is no such thing as impact-free energy today.

- Every time energy is converted from one form to another there is inefficiency; in other words, energy is lost. In many cases, that lost energy takes the form of waste heat or other forms of radiation, which have impacts of their own.

Globally, the demand for energy continues to rise.[1] We know what that means to consumers: rising prices for fuels such as gasoline and natural gas, as well as rising prices, shortages, and occasional brownouts or "rolling blackouts" for electricity. In addition, we're increasingly aware of the impacts of our energy usage, particularly in the form of GHG emissions.

In an ideal world, we'd have a readily available, zero-impact, safe and cost-effective source of energy that would meet our growing needs. Unfortunately, no such energy source exists today. So, we're left with taking a two-pronged approach: continuing to innovate and put into production new, lower-impact energy sources, while at the same time attempting to dramatically decrease our consumption through increased efficiency and changes of behavior. Of course, we could stop traveling, going to work, surfing the Web, shopping, and so on—and some people would advocate that we need to do that—but most of us would prefer to maintain our current lifestyle, so we need to make some serious headway on these two parallel energy strategies.

To engineers this means we need to wring every last bit of energy out of the products we design. The good news is that efficiency is a true win-win-win situation. If you can make your product using less energy, you can make it for less money. If your product uses less energy to operate, you save money for your customers. And in both cases, the environment wins as well.

Common Sources of Energy

Each type of energy has a different set of impacts. Some require mining or drilling; most require specialized equipment of some kind; and each has a different type and amount of emissions. Let's take a quick look at the most common energy sources today.

- Coal: Of the fossil fuels, coal is the most carbon-intensive and environmentally challenging to produce and use. Coal is used for electricity production in countries where the economies of large-scale consumption permit it to be used at low cost and in compliance with regulations on air and other emissions. Burning coal naturally produces very large volumes of CO_2; the mining of coal has undesirable impacts as well.

- Petroleum: Petroleum is relatively easy to transport and use and has an intermediate level of carbon intensity. The proportion of hydrocarbons ranges from as much as 97% by weight in the lighter oils to as little as 50% in the heavier oils and bitumens. Petroleum can be produced from conventional resources with relatively small environmental effects, and is readily available in international markets.

- Natural gas: Like petroleum, natural gas is currently abundant in large quantities, but it can be difficult and carbon-intensive to move it to consumers. Also, due to the need to cool natural gas to liquefy it, the cost of liquefied natural gas is typically an order of magnitude greater than that of the gas itself.

- Propane:[2] Propane is a byproduct of natural gas processing and petroleum refining. Volatiles such as butane, propane, and large amounts of ethane are removed from the raw gas to prevent condensation in natural gas pipelines. Oil refineries also produce some propane as a byproduct of production of cracking petroleum into gasoline or heating oil. When sold as fuel, it is commonly known as liquefied petroleum gas (LPG or LP-gas), which can be a mixture of propane along with small amounts of propylene, butane, and butylene.

- Uranium: Uranium is the primary mineral used as a nuclear fission fuel. Because of the slow growth of nuclear power worldwide, demand for a second fuel source has not yet arisen. By the time it is completely fissioned, one kilogram of uranium-235 can theoretically produce about 20 trillion joules of energy (20×10^{12} joules); as much electricity as 1,500 tons of coal.[3] Uranium is found in many countries but is produced in only a few. Worldwide, conventional uranium resources can meet current demand for at least a century, and even longer if estimated additional resources are taken into account.

- Biomass: Biomass includes all living plant matter as well as organic wastes derived from plants, humans, marine life, and animals. Trees, grasses, animal dung, as well as sewage, garbage, wood construction residues, and other components of municipal solid waste are examples of biomass. In the past, biomass was the primary source of fuel for the world. Today, in many developing countries it remains an important energy source for heating and cooking. As concerns about our increasing use of fossil fuels mount, many developed countries are now reexamining the potential for biomass energy to displace some use of fossil fuels. However, its higher costs compared with fossil and nuclear-based alternatives continue to handicap its growth in developing countries.

- **Geothermal energy:** In general, geothermal energy is thermal energy stored within the Earth's crust. Its big advantage is that it's a heat source that doesn't require the purchase of fossil fuels; however, thermal energy is not always easy to extract. From an economic perspective, geothermal energy is price-competitive with fossil fuels, but from an engineering perspective it can be inefficient since much of the heat energy is lost on extraction.

- **Hydroelectric power (hydropower):** Hydropower captures the stored energy in water that flows from a higher to a lower elevation under the influence of gravity. It produces virtually no harmful emissions and is not a significant contributor to global warming. Hydropower can also be far less expensive than electricity generated from fossil fuels or nuclear energy. Typically, hydroenergy conversion can be very efficient, with installations ranging in scale from a few kilowatts to more than 10,000 megawatts. Hydropower accounted for 6.4% of total U.S. electricity generated in 2005, according to the U.S. Department of Energy.[4]

- **Solar energy:** Solar energy is radiant energy from the sun. Sunlight is converted into electricity via photovoltaic systems or experimental technologies such as thermoelectric converters, solar chimneys, or solar ponds. The EPIA/Greenpeace Advanced Scenario shows that by the year 2030, photovoltaic (PV) systems could be generating approximately 2,600 TWh (terawatt hours) of electricity around the world. This means that, assuming a serious commitment is made to energy efficiency, enough solar power would be produced globally in 25 years to satisfy the electricity needs of almost 14% of the world's population.[5]

- **Wind energy:** Wind energy is an abundant, clean energy source that can reduce GHG emissions when it displaces electricity derived from fossil fuels. At the end of 2007, worldwide capacity of wind-powered generators was 94.1 gigawatts, according to the Global Wind Energy Council News. Most wind power is generated by wind turbines. Large-scale wind farms are connected to electrical grids; individual turbines can provide electricity to isolated locations.

Electricity and **hydrogen** are **secondary sources** of energy, which means they are used to store, move, and deliver energy in easily usable form. Today the cheapest way to get hydrogen is to separate it from natural gas, a nonrenewable energy source. Hydrogen can also be separated from water and from renewables, but hydrogen made from these sources is currently too expensive to compete with other fuels. Figure 7-1 summarizes consumption of primary and secondary sources of energy in the United States.

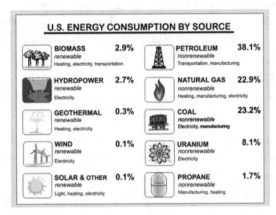

FIGURE 7-1   Sources of Energy Consumed in the United States (Data from the U.S. Department of Energy)

Calculating Energy and Power[6]

Many people (mistakenly) use the terms *energy* and *power* interchangeably. Energy is the ability to do work, and power is the rate at which work is done. To do 100 joules of work, you must expend 100 joules of energy. Power is the rate of using energy, so if you do 100 joules of work in one second (using 100 joules of energy), the power is 100 watts.

Over time, many different units for measuring energy have been developed. Table 7-1 shows some of the most common energy units and their mathematical conversion factors. If you want to convert units of fuel commonly used in the energy trade, such as barrels of oil, into calories or joules or British Thermal Units (BTUs), you simply multiply by the factors shown in the appropriate row (e.g., 1 BTU = 252 calories).

Note that since the chemical composition of some resources is variable with fuel source and extent of purification, the heat contents per unit mass or volume of these fuels also vary. For example, the heat content of 1.0 cubic foot of natural gas can range from 950 to 1,200 BTUs. For many engineering calculations, this provides sufficient accuracy for "orientation" or "scoping" calculations. However, keep in mind that the amount of useful energy obtainable from a given quantity of fuel can depend on how it is processed and utilized, and will always be less than the total energy content of the fuel.

First, a quick recap of the energy units used in the table is in order.

- **BTUs (British Thermal Units):** BTU is a unit of energy commonly used in the power, steam generation, heating, and air conditioning industries. In scientific contexts, the BTU has largely been replaced by the joule: 1 BTU = 1,055.05585 joules.

- **Joules:** Joules is the International System of Units (SI) unit of energy. One joule is the work done, or energy expended, by a force of one newton moving 1 meter along the direction of the force.

- **Quads:** A quad is a unit of energy equal to 10_1^5 (a short-scale quadrillion) BTU, or 1.055×10^{18} joules (1.055 exajoules or EJ). This unit is most commonly used by the U.S. Department of Energy in discussing world and national energy budgets.

- **Calories:** The calorie is a unit of heat. In most fields its use is archaic, and the joule is more widely used. However, it remains in common use as a unit of food energy: 1 calorie = 4.18400 joules.

- **Kilowatt-hours (kWh):** One kilowatt-hour is exactly 3.6 megajoules, and is the amount of energy transferred if work is done at a rate of 1,000 watts for one hour. This unit is typically used by electric utilities to express and charge for energy delivered.

- **Megawatt-years (MWy):** This is a variation on kilowatt-hours. One MWy is 8.76×10^6 kWh.

- **Barrels of oil (bbls):** A barrel of oil = 158.987295 liters.

- **Metric tonnes (tons) of oil:** The amount of energy released by burning one metric tonne (ton) of crude oil is approximately 42 gigajoules (GJ).

- **Metric tonnes of coal:** Electric power plants need about 1 kg of coal to produce around 2,000 watt-hours of electrical energy.

- **Thousand cubic feet (MCF) gas:** This is a unit of measure typically used in the oil and gas industry for natural gas.

- **Exajoules (EJ):** This is the SI unit of energy equal to 10^{18} joules.

TABLE 7-1   Various Energy Units and Conversion Factors

	BTUs	CALORIES	KWH	MWY	JOULES
BTUs	1	252	2.93×10^{-4}	3.35×10^{-11}	1,055
Quads	10^{15}	2.52×10^{17}	2.93×10^{11}	3.35×10^{4}	1.06×10^{18}
Calories	3.97×10^{-3}	1	1.16×10^{-6}	1.33×10^{-13}	4.19
kWh	3,413	8.60×10^{5}	1	1.14×10^{-7}	3.6×10^{6}
MWy	2.99×10^{10}	7.53×10^{12}	8.76×10^{6}	1	3.15×10^{13}
Bbls oil	5.50×10^{6}	1.38×10^{9}	1,612	1.84×10^{-4}	5.80×10^{9}
Metric tonnes oil	4.04×10^{7}	1.02×10^{10}	1.18×10^{4}	1.35×10^{-3}	4.26×10^{10}
Kg coal	2.78×10^{4}	7×10^{6}	8.14	9.29×10^{-7}	2.93×10^{7}
Metric tonnes coal	2.78×10^{7}	7×10^{9}	8,139	9.29×10^{-4}	2.93×10^{10}
MCF gas	10^{6}	2.52×10^{8}	293	3.35×10^{-5}	1.06×10^{9}
Joules	9.48×10^{-4}	0.239	2.78×10^{-7}	3.17×10^{-14}	1
EJ	9.48×10^{14}	2.39×10^{17}	2.78×10^{11}	3.17×10^{4}	10^{18}

Note: To convert from the first column of units to other units, multiply by the factors shown in the appropriate row (e.g., 1 BTU = 252 calories). Assumed caloric values: oil = 10,180 cal/g; coal = 7,000 cal/g; gas = 1,000 BTU/ft^3 at standard conditions.

Energy Impacts: Finding the Cleanest Source of Power

You may not have considered it, but the power that comes out of your wall socket may have a different set of environmental impacts than the power on the other side of town, or from another state, or from another country.[7]

If your power comes from coal, for example, it causes more GHG emissions than power derived from hydroelectric, nuclear, wind, or solar sources. That's why the CO_2 emissions associated with power generation are higher in Colorado than in California; and it's why France generates near-zero CO_2 for the energy it produces with its large bases of nuclear power plants. And as we've pointed out, these energy sources with lower GHG emissions aren't without their own impacts.

As an engineer, you can't control where your customer uses the product you've designed and built, but you still have an important role to play. You may be able to influence the types of power your suppliers use in manufacturing parts for your product. You can also choose less impactful forms of transportation. You can provide your customers with options to use cleaner energy sources where possible. And, of course, reducing energy usage by design helps avoid the whole discussion.

As an example, one option that corporations with large data centers are now exploring is referred to as "chasing the sun." With this approach, power supplied to data centers is sourced from the cleanest available energy, which is typically being generated from solar panels or wind turbines during daylight hours. Through arrangements with power providers, a data center in Washington, D.C., could be powered from solar cells in California. In theory, this concept could also be applied internationally so that a New York data center could take advantage of a cloudless day in Australia or a company in Japan could take advantage of an especially windy day in Palm Springs or the Sierra foothills—constantly maximizing its use of the greenest energy available.

ENERGY COST OF A SHIRT

Here are some facts courtesy of Patagonia CEO and founder, Yvon Chouinard.[8]

- From raw materials, it costs 110,000 BTUs to make a Patagonia shirt.
- Shipping it airfreight from Ventura, California, to a store in Boston costs 50,000 BTUs.

Generically, the cost to ship per ton is as follows:

- Rail or boat: 400 BTUs per ton per mile
- Truck: 3,300 BTUs per ton per mile
- Air cargo: 21,760 BTUs per ton per mile

Shipping shirts to store shelves via airfreight can increase the total energy expense by almost 50 percent.

Energy and GHG Emissions

The preceding section discussed the connection between energy use and CO_2 emissions, and the need to measure and minimize both throughout the product lifecycle. But there are many other sources of emissions and waste. This section summarizes the key sources of GHG and other emissions, and provides information about measuring and minimizing their impact.

Greenhouse Gas Primer

Although we often discuss GHG emissions as "tons of carbon dioxide" or even less accurately as "tons of carbon," there's more to GHG than CO_2. According to the U.S. Environmental Protection Agency (EPA), the principal GHGs that enter the atmosphere because of human activities are as follows.[9]

- **Carbon dioxide (CO_2):** Carbon dioxide enters the atmosphere when fossil fuels are burned; it is "sequestered" or removed from the atmosphere when it is absorbed by plants.

- **Methane (CH_4):** Methane emissions come from livestock, the decay of organic waste in landfills, and the production and transport of coal, natural gas, and oil.

- **Nitrous oxide (N_2O):** Many different types of agricultural and industrial activities generate nitrous oxide; it is also produced during combustion of fossil fuels and solid waste.

- **Fluorinated gases:** These GHGs are typically emitted in small but potent quantities; they are sometimes referred to as High Global Warming Potential gases. Hydrofluorocarbons, perfluorocarbons, and sulfur hexafluoride are examples of synthetic GHGs that are emitted from industrial processes.

CO₂ Equivalents and Conversions

GHGs contribute to global warming in varying degrees. The term *carbon dioxide equivalent* (CO_2e) provides a standard way to calibrate the global warming potential (GWP) of various gases; specifically, CO_2e is the amount of carbon dioxide that would yield the same warming effect as a particular greenhouse gas or greenhouse gases. It is used in carbon accounting—for example, to account for GHG emissions and reductions over time.

On the scale supported by the Kyoto Protocol, CO_2 is the reference point and has a GWP of 1. Every other GHG has a greater GWP than CO_2. Table 7–2 provides the conversion rates.[10]

TABLE 7–2 CO_2e Conversion Factors

GHG	MULTIPLY BY THE FOLLOWING FIGURE TO OBTAIN THE CO_2E VALUE:
CO_2	1
CH_4	23
N_2O	296
SF_6	22,200
HFCs	12–12,000
PFCs	5,700–11,900

(Source: Third Assessment IPCC Report, 2001)

Calculating GHG Emissions

GHG emissions are usually fairly straightforward to calculate. If you know the specific source of the emissions, there are well-documented formulas for arriving at the resulting amount. For example, burning a gallon of gasoline in a car results in 19.9 pounds of CO_2e emissions (you may be objecting, pointing out that a gallon of gas weighs less than 19.9 pounds, but the extra weight comes from the addition of oxygen in the air during the combustion process). Table 7-3, compiled by the United Kingdom's Department for Environment, Food and Rural Affairs (Defra), provides conversion factors for various fuel types.[11]

In most situations, the standard units of reporting GHG emissions are metric tons of CO_2e. Obviously, tons are cumbersome when dealing with very large or very small amounts of emissions, but they are a good fit for many situations.

In some cases, we don't know the specific source of the emissions. For example, the electricity delivered to your home comes from a wide range of sources. In these situations, you can attempt to get a specific number from your electric company, but that may not always be possible or practical. Fortunately, there are accepted accounting standards for many common situations that you can draw on. For example, if you're interested in the impacts of electricity in the United States and you can't get a number from the source, the U.S. EPA Emissions & Generation Resource Integrated Database (eGRID)[12] is a widely accepted source for calculations.

The eGRID numbers are also good if you are estimating the impact of power use at a customer or manufacturing partner and don't know the source of their electricity. Here are a few interesting facts and examples from eGRID.[13] In the United States:

- 12,100 pounds of CO_2e emissions from gasoline is about average for one vehicle over a year (assuming an average of 231 miles/week per vehicle).

- 11,000 pounds of CO_2e emissions from natural gas use is average for a household of two people over a year.

- 16,290 pounds of CO_2e emissions from electricity usage is about average for a household of two people over a year.

If all of this sounds a little like corporate accounting, it is. A GHG inventory is a formal accounting of the amount of GHGs emitted to or removed from the atmosphere over a specific period of time (e.g., one year) from a defined set of activities. A GHG inventory also provides information on the

TABLE 7-3   Fuel Conversion Factors

Converting fuel types to CO_2			Net CV Basis
Fuel Type	Units	x	kg CO_2 per unit
Electricity	See Annex 3		
Natural Gas	kWh	x	0.206
	therms	x	6.023
Gas Oil	tonnes	x	3190
	kWh	x	0.265
	litres	x	2.674
Diesel	tonnes	x	3164
	kWh	x	0.263
	litres	x	2.630
Petrol	tonnes	x	3135
	kWh	x	0.253
	litres	x	2.315
Fuel Oil	tonnes	x	3223
	kWh	x	0.281
Burning Oil	tonnes	x	3150
	kWh	x	0.258
	litres	x	2.518
Coal	tonnes	x	2457
	kWh	x	0.346
LPG	kWh	x	0.225
	therms	x	6.608
	litres	x	1.498
Coking Coal	tonnes	x	2810
	kWh	x	0.349
Aviation Spirit	tonnes	x	3128
	kWh	x	0.250
	litres	x	2.233
Aviation Turbine Fuel	tonnes	x	3150
	kWh	x	0.258
	litres	x	2.518
Other Petroleum Gas	tonnes	x	2894
	kWh	x	0.217
Naphtha	tonnes	x	3131
	kWh	x	0.249
Lubricants	tonnes	x	3171
	kWh	x	0.263
Petroleum Coke	tonnes	x	3410
	kWh	x	0.361
Refinery Miscellaneous	kWh	x	0.259
	therms	x	7.585
Total			

Note: Carbon emissions are usually quoted in kg CO_2/kWh. If you wish to convert the carbon dioxide factors into carbon equivalents (i.e., kgC/kWh), multiply the figure by 12 and divide by 44.

activities that cause emissions and removals, as well as background on the methods used to make the calculations.

A thorough discussion of GHG accounting and the WRI/WBCSD Greenhouse Gas Protocol (the most widely used international accounting tool), along with tools for calculating direct and indirect GHG emissions, is available at www.ghgprotocol.org.

The GHG Protocol's **Corporate Standard** (see www.ghgprotocol.org/standards/corporate-standard) provides standards and guidance for companies and other organizations preparing a GHG emissions inventory. It covers the accounting and reporting of the GHGs covered by the Kyoto Protocol, and was designed to help companies prepare a GHG inventory that represents a true and fair account of their emissions; to simplify and reduce the costs of compiling a GHG inventory; to provide information that can be used to build an effective strategy to manage and reduce GHG emissions; and to increase consistency and transparency in GHG accounting and reporting.

The Greenhouse Gas Protocol Initiative also provides a range of calculation tools for GHG emissions and answers specific questions about GHG accounting. The tools and guidance are available at www.ghgprotocol.org/calculation-tools.

Putting a Value on Carbon (Dioxide!)

In recent years **carbon trading** markets have emerged to place economic incentives on reducing GHG emissions. For example, in a **cap and trade** program, a central authority (usually a government) sets a limit or *cap* on the amount of a pollutant that can be emitted; companies or other groups are then issued emission permits and are required to hold an equivalent number of allowances or credits that represent the right to emit a specific amount. The total amount of allowances and credits cannot exceed the cap, limiting total emissions to that level. Companies that need to increase their emissions must buy credits from those that pollute less.

With lots of buyers and sellers, a market has emerged for the right to emit a certain amount of GHG. In effect, the buyer is paying a charge for polluting, and the seller is being rewarded for having reduced emissions by more than was needed.* Thus, in theory, those that can easily reduce emissions most

---

* The **Acid Rain Program** undertaken by the U.S. EPA provides a good example. In an effort to reduce overall atmospheric levels of sulfur dioxide and nitrogen oxides, which cause acid rain, the program implemented emissions trading primarily targeting coal-burning power plants, allowing them to buy and sell emission permits (called "allowances") according to individual needs and costs. For details see www.epa.gov/acidrain/index.html.

cheaply will do so, achieving the pollution reduction at the lowest possible cost to society.[14]

There are many ways to implement a cap and trade system, and we suspect that we're all going to be hearing more about them over the coming years. Other proposals involve a **carbon tax,** whereby emitters would have to pay the government a set amount. In places such as the United States where there is no mandatory GHG market or tax, some companies voluntarily participate in carbon trading to support internal goals such as carbon neutrality (see some additional thoughts on carbon neutrality in Chapter 11).

An important result of any of these systems is that companies that are involved (by law or voluntarily) can now put a real financial value on the right to emit GHGs. Rather than discussing the soft-dollar benefits and potential savings of a product, companies can perform a true ROI analysis that includes hard-dollar carbon savings.

As an engineer, it is important to know whether your company, partners, or customers have established a "price for carbon" for their internal use. If they have, you will find that the ROI analysis related to energy efficiency will now reflect an increased savings due to the decrease in GHG emissions that will accompany decreases in energy use. As a result, some projects that could not clear the financial hurdle based only on energy savings may now be able to clear it and get the green light.

Heat, Noise, Light, and Radio Emissions

Products that use energy give off emissions beyond GHGs. Each of the following categories is actually worth a book in itself, but here are just a few basic guidelines.

- **Heat:** Heat is a wasteful side effect of just about every energy-using product. How long will your car run without engine coolant? How long are you comfortable working with your laptop on your lap before your legs get uncomfortable? Waste heat can be dangerous to both humans and the products themselves. And not only does heat represent wasted energy, but some products—such as computers and cars—have to expend even more energy to help cool themselves. So, from an environmental perspective, a financial perspective, and a human perspective, engineers need to find new ways to minimize waste heat. Choosing components carefully and being willing to pay a price premium for cool-running and low-noise characteristics can help a lot.

No single book or reference source covers every aspect of this topic for the vast spectrum of device types being made today, but you can get many broadly applicable pointers from *Building the Perfect PC*, Second Edition, by Robert Bruce Thompson and Barbara Fritchman Thompson.[15]

- **Noise:** Unless your product plays music, is a growling roadster, or needs to alert people, product noise is generally not a good thing. Since sound waves are just another form of energy, unwanted noise represents wasted energy and can be a sign of mechanical or electrical inefficiency. It's often a sign of components that might be under additional stress and may be liable to fail. But the impact goes beyond inefficiency: Unintended noise can represent a human health hazard or create unexpected environmental effects—or even disrupt migratory patterns. Unintentional noise from a product is a sign of mechanical or electrical inefficiencies. That's why there are laws relating to product noise. In the United States, state and local laws will apply, but depending on where your product is going to go you may need to familiarize yourself with international noise laws as well.

- **Electromagnetic interference:** Electromagnetic radiation from products, including light and radio emissions, can result in electromagnetic interference (EMI, also known as radio frequency interference when it is isolated to the radio frequency bands), and can, in some cases, be directly dangerous to humans (hence the lead vest when you get your dental X-rays). Standards are in place to specify which EMI levels are acceptable for various products under various conditions. The most common EMI testing on electronic equipment for the United States is FCC Part 15 testing. FCC Part 15 covers unintentional testing and evaluation as well as low-power unlicensed transmitters. You can find more information on this at www.cclab.com/fcc-part-15.htm. In Japan, the EMI testing standard is outlined by the VCCI. More information about VCCI testing is available at www.cclab.com/vcci.htm.

Process-Related GHG Emissions

While we've focused on energy as the main source of GHG emissions, GHGs are also emitted from manufacturing processes. High-GWP gases are emitted from a variety of industrial processes, including aluminum production, semiconductor manufacturing, electric power transmission, and magnesium production and processing. In addition, some high-GWP gases are being used to replace ozone-depleting substitutes such as chlorofluorocarbons (CFCs).

This is yet another advantage of recycling, as the process difference can also result in emissions differences. For example, high-GWP gases are emitted during primary aluminum production, not aluminum recycling.

Energy Efficiency in Product Design

For products that use energy in operation, one of the most impactful things to do as an engineer is to improve the energy efficiency. Here are a few specific topics to consider.

Core Efficiency

In most cases, the bulk of the power a product uses during operation is being converted into core functionality. The car is converting gasoline into mechanical power that can spin the wheels. The smartphone is converting stored electricity into processing, transmitting, and receiving radio waves, illuminating a screen, and so forth. An oven is converting gas or electricity into heat.

By improving core efficiency, you can reduce the cost of your product for the customer—but equally important, you can create other positive side effects. For example, using less energy means less conversion loss (discussed shortly). Lower energy requirements can enable you to use smaller batteries or other components, resulting in lower weight and decreased manufacturing cost and impact. And less energy usually means less waste heat, which can result in lower power requirements for fans or air conditioning.

Every product design effort should spend time setting goals regarding core efficiency, and invest in engineering and innovation to optimize this important area. The direct benefits can be huge, and can often result in product improvements in other areas as well.

Energy Transmission and Conversion

Every time energy is converted or moved, we lose some of it. When we convert electricity from AC to DC or from 12V to 5V, we lose energy in the process. When we convert gas to motion, we lose energy. When we convert electricity to chemical energy stored in a battery, we lose energy. When we transmit energy from one region to another, from one room to another, or even from one side of a silicon chip to another, we lose energy.

A few years ago it was not uncommon for a computer to waste 30% to 40% of the power that it consumed during the process of converting from alternating current (AC) that comes from the wall to direct current (DC) that computer electronics use. With recent improvements most manufacturers have now reduced those losses to 20% or less, but these are still significant losses.

There are three ways to attack this problem. First, you can reduce the energy that the product uses. If you need to convert less energy, you will lose less in the process. Second, you can focus on the core transmission or conversion process. This may be as easy as selecting a different component, or it may require some fundamental design changes. Pay special attention to ensure that conversion components are not oversized. Many types of conversion have a sweet spot where the percentage loss is minimized, and operating way outside of that can result in higher waste. Finally, you can look at the overall system and try to eliminate unnecessary conversions.

Power States

Increasingly, product designers are using the concept of **power states** to reduce the total power used by a product. Power states are different modes that products can switch between, either automatically or under user control. One of the most common examples is your home computer, which has separate power states for the monitor and processing unit. For cars, Chevy now has a motor for some of its trucks that recognizes when the vehicle is being used in a highway situation, and automatically converts from a V-8 to a more efficient V-4 engine by turning off four of the pistons. Your cell phone will blank its screen if you haven't touched it for a while.

"The key is to ensure that a product's energy consumption is proportional to the work it's doing," says Sun engineer Subodh Bapat. "Today, unfortunately, that's not the case with most products. And the reason boils down to bad design. In many cases, a small chip and a bit of code could solve the problem entirely."

Since power states usually have some impact on the functionality of the product, it is important that you give users some control over when automatic state transitions occur. And it's important to do some work to try to get a good "out of the box" setting—the design work you do on power states will be worthless if users get fed up and just turn them off.

Standby Power

One common power state, standby power, is not as eco-friendly as it first appears. This is because consumers are led to believe that the product is in a low-power state, or in many cases even totally off, when in fact the product is still drawing a not insignificant amount of electricity. And since these are common products in many countries, the total amount is adding up.

Televisions, for example, use standby power to save settings and to make the TV turn on quicker. But that standby power amounts to 10% to 15% of the total energy use on average. Even when the TV is "off" it's on, drawing 15–20 watts. Microwave ovens, game consoles, DVD players, and digital cable boxes all draw 20–25 watts of power when they're not being used. Desktop computers and laptops have historically drawn about 25 watts even while in sleep mode.[16]

Fortunately, engineers are now beginning to improve product design to reduce even standby power consumption. For example, some of the newer PCs will typically draw only about 2–3 watts of power while they sleep, which is a big improvement.[17]

Solutions for consumers are also emerging. You can now purchase special power strips (called Smart Strips or Watt Stoppers) that monitor power consumption, sense when devices are using standby power, and turn them off. Usage monitors measure precisely how much power is being used (or wasted) by a given device at a given time. And many lighting systems now detect the lack of people in a room and will turn off the lights.

Batteries

Batteries play an important role in a wide range of products. Some are obvious, such as watches and mobile personal electronics (iPods, cell phones, etc.). Others are hidden, overshadowed by another primary energy source. Cars use batteries to start and maintain some functions when the engine is off; PCs and electric clocks use batteries to remember the time when electricity isn't available; video games and pinball machines use batteries to remember high scores.

One of the important decisions in product design concerns whether to use rechargeable or disposable batteries. Disposable batteries are convenient, but lose effectiveness over time and have disposal issues. Rechargeable batteries can be used for a much longer period of time, but still have disposal issues, can be inconvenient (users need to carry a charger and remember to charge them), and can lose effectiveness over time. In addition, chargers are often inefficient, wasting energy when they have completed charging up the battery.

This last point has become an issue, especially for large-volume products such as cell phones, where billions are in use today. Most people will plug them in for long periods of time to recharge, unaware that the recharger continues to draw some power even after the battery is charged.

In the end, different products take different approaches. Apple, for example, went with a built-in battery in its iPod designs, thereby simplifying both the use of the product for customers and the take-back of batteries because Apple does the battery replacement as a service. We suspect that Apple realized that most consumers will get a new iPod before they'll need a new battery. Cell phones also typically use built-in batteries today, but ones that are easy for consumers to replace. This means it is easier for consumers to replace a bad battery, but stores now have to stock huge arrays of replacement batteries and it's harder to ensure that the dead batteries are recycled correctly. Some products are better off with disposable batteries; for example, a GPS device for hikers that's used only a few times each year. It's easier to throw a few spare AA batteries into your backpack than it is to go out and buy one or more expensive, product-specific replacement batteries that need to be charged.

In short, deciding whether to use batteries and deciding on the type of battery to use can be important decisions for product designers, and shouldn't be taken lightly. If your product is going to use batteries, try to use the smallest possible battery. If there is an option for avoiding battery use, explore it.

Tracking Lost Energy

As you model efficiency and look at the places that energy is lost in your product, it's important to ask yourself where the lost energy is going. Is it turning into heat that could injure a user, require fans or air conditioning, or increase the failure rate of the product? Is it turning into electromagnetic radiation that could cause issues with meeting key product requirements? Is it being lost in mechanical systems and resulting in excess wear on parts?

The unintended consequences of lost energy can significantly impact not only the total footprint of your product, but also its viability in the market. Make sure you fully understand and account for the energy use of your product.

An Example: Energy Efficiency in Data Centers

Data centers full of servers and data storage systems are at the heart of our modern economy. In developed countries, every economic transaction is recorded at least once, and often in multiple places. Just think about how much electronics is involved when you perform a simple search on Google: your desktop, the network in your house, the network to Google (which goes through multiple large net-centric data centers), and the servers and storage in the data centers at Google (which, the search results tell us, is not insubstantial). Then when you click on a result all of those paths are used again, but this time to go to another data center that houses the site you want to visit. Every email, medical procedure, phone call, paycheck, online purchase, plane ticket, and electronic toll payment involves one or more data centers.

Power consumption of data centers doubled between 2000 and 2005, and is expected to double again by 2010. For the entire world, doubling server consumption from 2005 to 2010 would require additional capacity equal to more than ten additional 1,000-megawatt power plants.[18] And today, according to analyst firm IDC, roughly 50 cents is spent on energy for every dollar of computer hardware—this is expected to increase by 54% to 71 cents over the next four years.[19]

So, although data centers represent a small fraction of global GHG emissions (estimated to be less than 2%), their rapid growth and the economic impacts of their energy use are attracting attention. Even Congress has gotten into the act with Public Law 109–431 (enacted in 2007), "to study and promote the use of energy-efficient computer servers in the United States."

Where Energy Goes in Data Centers

It's not just the servers and storage systems that consume energy. Often, the equipment that is required to cool the servers and the server room uses as much power as the systems themselves. Add to that the energy used to light the data center, the power distribution loss, and other factors, and you'll see that the majority of power coming into the data center is used for something other than IT equipment.

Figure 7-2 illustrates this point.[20] Where the energy goes in any individual data center depends upon the age and efficiency of the equipment, adherence

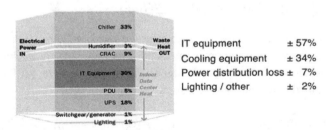

FIGURE 7-2   Use of Power in a Typical Data Center

to best practices, levels of redundancy, and so on, but this graphic shows why IT equipment often consumes less than half the power used in a typical data center. It also drives home the point that the overall energy system is far bigger than the energy-consuming devices themselves.

Let's start at the bottom of the picture and work our way up. First we have to light the room, since humans sometimes have to work there. Next, a large electric feed from the utility enters the building and is switched. Since many data centers are business (or mission) critical, and since power from the utility is not fully reliable, we put in an uninterruptible power supply (UPS) as a backup in case of power supply glitches. Finally, the power passes through power distribution units (PDUs), where it is distributed out to racks of equipment.

The equipment (networking, servers, storage) takes the energy in, converts it to different voltages (losing some energy in the process), and powers the components. Since very little physical activity is involved (spinning disks and fans, mostly), the bulk of the power ends up converted to heat (with a little noise and mechanical vibration thrown in). Also note that the PDUs and UPS are not totally efficient, so they generate some heat as well.

Getting rid of megawatts worth of heat day-in and day-out is not easy. First there are computer room air conditioners (CRACs), which capture the heat and get it out of the room. That heat is then finally taken care of by industrial chillers (these often use fresh water and have become a separate target for eco improvement). Finally, you may have noticed that 3% or so of the power is used for humidifiers. Why, you might ask? The air conditioning process removes moisture from the air, and eventually the air can get too dry. As a result, we need to add a step to put some of the moisture back into the air.

Amazingly, when you add this up, the IT equipment itself is using only about one-third of the energy, but *other than the lights, all of the remaining energy is there just to support the delivery and removal of that energy.* In other words, every watt of IT equipment in this scenario requires two more watts just to process it.

Making Data Centers More Efficient

With rising energy prices, the energy usage by these large data centers has not gone unnoticed. Companies throughout the industry are working individually, and in groups such as the Green Grid, to try to drive out inefficiencies throughout the system. Here are some examples.

- **More efficient products:** As we've seen, if you can use one less watt in IT equipment, you will often save one or more additional watts that would have been lost in delivery and used for cooling. Advances such as chip multithreading (CMT), slower disk drives, and automated power-down technology are beginning to be widely used with good results.

- **Rating systems for equipment purchasers:** While eco-rating schemes such as Energy Star, 80 Plus, EPEAT, and Climate Savers Computing have long been available for desktops, standard ratings for servers tend to be more complicated, so they have lagged behind. In the meantime, Sun and other companies have devised increasingly accurate power calculators that help you compute the expected power consumption of specific server configurations. These are useful for two reasons. First, they allow some head-to-head comparisons (recognizing that you're trying vendors' own numbers). Second, they can help customers to size infrastructure equipment such as PDUs and CRACs, since oversizing those systems can lead to very large inefficiencies.

- **Consolidation:** Most existing servers use most of their peak power whether they are 20% loaded or 80% loaded. As a result, you can realize big savings by combining jobs onto the same system, which ideally you can keep at a higher utilization level. Virtualization technology, which enables different jobs to run safely on the same system, is one of the most dynamic areas in computing today.

- **Conversion/transmission efficiency:** A recent study by Lawrence Berkeley Labs and Sun examines new ways to obtain more efficient high-voltage power distribution, and compares high-voltage AC and DC and the potential energy savings of each. "The first order estimates are that high-voltage DC could save 5% to 7% over high-voltage AC, but the big savings is going to high voltage in general," says Hal Stern, vice president of Systems Engineering at Sun. Customers and vendors are also looking to minimize the number of power conversions that occur from the utility feed to the equipment.

- **Power states:** You generally don't want your servers to put themselves totally to sleep, but many new kinds of power states are being experimented with in upcoming server and storage designs.

- **Creative cooling:** Cooling is suddenly an area of major innovation, with customers and vendors exploring new options such as integrating cooling technology directly into server racks, using variable-speed fans rather than single-speed fans, and making better use of cool external air (a.k.a. "fresh air cooling").

- **Code efficiency:** Less code, executed more efficiently, means fewer CPU cycles expended in processing workloads. For large compute farms, a 10% improvement in efficiency can mean 10% fewer machines and 10% less energy.

- **Technology refresh:** Computing technology continues to follow an exponential improvement curve. Combine that with new features such as power states and upgrading older equipment and you can often save enough energy to fund the project.

Example Results

We can point to our own experience to illustrate how significant the results of energy-saving efforts can be: Sun recently built new energy-efficient data centers in the United Kingdom, India, and the United States, compressing a total of 152 data centers (202,000 square feet) into 14 new centers (76,000 square feet), resulting in a 60% reduction in overall power consumption—which cut utility bills by $860,000 in the first nine months. The new data centers also reduced new construction requirements at existing facilities, a cost avoidance of $9 million. The new facilities will reduce carbon emissions by 3,227 metric tons annually, according to Dean Nelson, Sun's director of Global Lab and Datacenter Design Services.[21]

8

Chemicals, Materials, and Waste

As *Cradle to Cradle* makes clear, everything that goes into a product eventually comes back in some form. And it's up to us, as engineers, to find more creative and eco-responsible ways to deal with it. Thinking of this from a lifecycle perspective, we need to consider the following.

- **The impact of sourcing specific materials:** Are the materials we're using dangerous to create? Are emissions related to their manufacture? Can we use recycled materials?

- **The safety aspects of products in use:** Does the product contain dangerous chemicals? Is there any potential for hazardous emissions?

- **The impact of materials at end-of-use:** Can the materials easily be taken apart and recycled? Are there any hazardous chemicals that will have to be reclaimed with special means (e.g., mercury in chlorofluorocarbons or CFLs)?

This chapter covers the regulations that impact the selection of materials, packaging and documentation considerations, and waste and renewal issues that impact product design such as disassembly, reusability, recycling, and take-back.

Chemistry and the Law

All of the potential dangers of chemicals and materials are not lost on the world's governing bodies. For example, the European Union's Reduction of

Hazardous Substances (EU RoHS) Directive already bans new electrical and electronic equipment from the EU market if it contains excessive levels of lead, cadmium, mercury, hexavalent chromium, polybrominated biphenyl (PBB), and polybrominated diphenyl ether (PBDE) flame retardants. But the RoHS Directive, passed in 2003, is only the tip of the iceberg.

Industrialized countries throughout North America, Europe, and Asia are now considering, drawing up, or in the process of implementing additional restrictions on the use of chemicals and materials. Current examples include the REACH (Registration, Evaluation, Authorization, and Restriction on Chemicals) regulations, dozens of new U.S. and EU battery laws, and a Chinese version of the RoHS Directive, among others. And multiple standards bodies worldwide are busy defining new requirements and standards for material declarations, based on Joint Industry Guide specifications, IPC data, IEC Working Group recommendations, and so on. It is important to note that these regulations not only include bans or limits on certain chemicals, but also involve reporting the usage of certain chemicals, or specialized handling of the chemicals at end-of-use of the product. Given the disparity of laws across different regulatory regions, these reporting and disposal regulations can become quite expensive to comply with.

Figure 8–1 provides a graphic example of how recent regulations have limited choices for engineers. When you eliminate from consideration the elements that are now banned for use in electronics by RoHS or other laws, the elements that may soon be banned through proposed legislation, and the elements that are useless in electronics, only about one-third of the periodic table remains available to engineers.

In addition, the WEEE Directive, made effective in July 2006, makes manufacturers responsible for e-waste, based on recovery, recycling, and collection targets. The legislation is designed with two aims in mind: to create an economic incentive for manufacturers to design more environmentally friendly products, and to reduce the environmental impact of waste by increasing the volume that is recovered and recycled.

To deal with this complexity, many companies are formalizing their product content restrictions through specification documents. Hewlett-Packard, for example, makes detailed specifications available to its supply chain covering product content and testing requirements for lithium and lithium-ion cells, batteries, and battery packs, as well as standards for the marking of plastic parts and products for subsequent decision making during the regeneration phase. If you work in a larger company, it's important to understand your company's standard approach to chemical decisions.

Useless in electronics. Gas at room temperature, or too rare, or too unstable, etc. Might be found in rare applications, like lasers or medical isotope imaging.

Banned by RoHS or other existing laws.

? Already restricted by some customers.

Leaves only one-third of the table!

FIGURE 8–1   Elements of the Periodic Table That Are Not Suitable for Manufacturing, Are Banned, or Are Restricted

For more detailed information about current regulations relating to materials and chemicals, see the following URLs:

- WEEE regulations:
  www.dti.gov.uk/innovation/sustainability/weee/page30269.html

- RoHS Directive: www.rohs.gov.uk/

- U.S. General Environmental, Health, and Safety (EHS) Guidelines, April 30, 2007:
  www.ifc.org/ifcext/enviro.nsf/Content/EnvironmentalGuidelines

- iNEMI standards: www.inemi.org/cms/

In addition, industry trade groups and consultants can often help you understand the legal implications and alternatives for materials decisions.

Packaging and Documentation

Take a look at any consumer product on a store shelf and you'll see why packaging is a growing concern for environmentalists. A simple electric razor, for example, is sold encased in a clear, rigid, molded plastic container (usually unopenable and virtually indestructible!) that houses a variety of separately packaged components: a cardboard box containing the razor blades, a power cord in a shrink-wrapped plastic tube; plastic-wrapped batteries; and another shrink-wrapped packet with various instructions and warranty cards.

Or consider a poorly packaged personal computer, which ships in a box of boxes—with a separate package for each component (even the power cord)—layered with molded polystyrene and cushioned by Styrofoam or hundreds of Styrofoam peanuts. And depending on who buys the product, these packaging materials may be headed straight for the landfill right after the goods are unpacked.

Two points here: First, packaging is often almost pure waste. As engineers, we need to stop asking ourselves how to make packaging more efficient and start asking how to get away with less of it. The good side of this is that, just as we have seen in other areas, there is the potential for some significant savings if we can figure out how to package products more effectively and efficiently.

Second, the "product" and the "packaging" have their own separate lifecycles and supply chains, and engineers who are designing for optimal environmental effectiveness need to consider both of them. Why? Regulations covering design and take-back of packaging materials are mushrooming throughout Europe, Asia, and North America, and compliance with these environmental packaging laws requires creative engineering.

Policies related to packaging began to spread rapidly after the EU Directive on Packaging and Packaging Waste was published in 1994. This measure spawned similar policies in Eastern Europe and eventually Asia, and today environmental packaging requirements apply to products sold in most global markets, including[1]

- The Americas (United States, Canada, Brazil, Mexico)

- Europe (EU member states, Norway, Iceland, Switzerland, Bulgaria, Croatia, Romania, Turkey, Ukraine)

- The Asia/Pacific region (Australia, China, Japan, Taiwan, South Korea, India, Bangladesh)

- Africa/the Middle East (Tunisia, South Africa, Israel)

The specific regulations for any given product tend to be in a perpetual state of flux. The challenge for engineers is to develop a packaging solution that will be marketable in as many regions as possible, while keeping the cost and complexity of compliance at a minimum. This requires a thorough understanding of the requirements in each jurisdiction.

For example, in the EU all packaging is subject to **Extended Producer Responsibility** (EPR) policies, meaning all components and complete packaging systems must be source-reduced, must comply with heavy-metals limits and minimization requirements for other noxious and hazardous substances, and must be recyclable, be compostable, and/or yield a certain energy gain when incinerated. The Packaging and Waste Directive[2] in Europe also mandates that companies selling products in Europe recover their product's packaging before it enters the waste stream. Many companies satisfy the requirements of the Directive by joining a "GreenDot" program (in Europe a total of 32 countries have national packaging compliance organizations that manage their country's packaging recovery programs).

Restrictions, bans, and phase-out limits also apply to certain materials, particularly expanded polystyrene (EPS) and polyvinyl chloride (PVC) in some jurisdictions and for certain product and packaging types. Several countries limit the percentage of empty space that may be contained within packaged consumer goods. Certain U.S. states require the use of recycled content materials in plastic packaging containers. In 2004, California's regulation was changed to require all rigid plastic packaging containers to comply regardless of the statewide recycling rate. And companies operating or trading in some markets must file periodic statements outlining packaging reduction efforts, goals, and progress toward existing goals.

In addition, there has been a recent surge in **take-back policies**, or regulations that require manufacturers to devise or fund a packaging recovery and recycling scheme. The Waste Electrical and Electronic Equipment (WEEE) and RoHS directives are the best-known sources of take-back regulations.

In many countries, fees are now imposed on packaged goods based primarily on the amount of packaging (by weight) and the type of packaging materials used. In general, the more packaging a product bears and the more difficult the packaging material is to recycle or manage in a given country, the higher the fees. Companies are required to calculate the quantities of each packaging material used and to file periodic reports—which, of course, requires detailed packaging data. For more detailed information about WEEE regulations and the RoHS Directive on take-back regulations, see www.dti.gov.uk/innovation/sustainability/weee/page30269.html and/or www.rohs.gov.uk/.

Clearly, engineers need to consider the full range of environmental issues related to packaging—across the full lifecycle of the product, across the entire supply chain—and be more creative in designing eco-effective packaging. But we must do so in a way that is economically practical, not just ecologically sensitive. We must find new ways to extract waste from the equation at the same time we devise better alternatives in terms of materials and production processes.

It is a common misconception that packaging design is typically an afterthought or a mundane chore for an engineer. Nothing could be further from the truth; in fact, the package design for many products can be far more ingenious than the product itself.

Consider the challenge of creating the packaging for Pop'n'Fresh dough. The mathematics involved in creating an airtight tube that will pop open easily when tapped against a hard surface—but not so easily that it will open prematurely in a refrigerator—would stagger many a rocket scientist. Even designing a cardboard box for shipping a refrigerator involves extremely complex calculations of "axial compression strength," formsboard options, paper grades, and post thicknesses to keep the refrigerator free from dents and scratches during shipment.

The point is that since packaging is already a sophisticated science, the time has come to apply a greater proportion of engineering ingenuity to eco-effective packaging. If we can optimize the axial compression strength of a refrigerator carton, surely we can reimagine the design, materials, and production processes to optimize for eco-effectiveness—throughout the product lifecycle, across the supply chain.

Waste and Renewal

The materials that go into product manufacture must be dealt with again when the useful life of the product is over. This section addresses the key considerations in designing for waste disposal and renewal: disassembly, reuse/recycling, and take-back.

Disassembly

In the past few years, a considerable amount of research has focused on designing products for disassembly and reusing/recycling their materials and components. Some of the key considerations for engineers include[3]

- Designing for easy disassembly and enabling the removal of parts without damage

- Labeling the parts so that people know what to do with them

- Ensuring that the purification process does not damage the environment

- Designing for ease of testing and classification

- Designing for reconditioning, or supporting the reprocessing of parts by providing additional material as well as gripping and adjusting features

- Providing easy reassembly for reconditioned and new parts

From the preceding list you see that disassembly plays an important role, not only in enabling parts and materials to be removed for recycling but also in enabling reconditioning, refurbishment, remanufacture, repair, and service of the product and components, extending their useful life.

Reuse/Recycling

Design for Recycling (DFR) initiatives are gaining momentum. It is becoming increasingly clear that it makes both economic and ecological sense to integrate end-of-life aspects into the design of products—particularly given the upward spiral in new legislation (e.g., packaging waste regulations such as the EU Packaging Directives, End of Life Vehicles [ELVs], and the WEEE Directive). Engineers are in a position to impact the recyclability of products because they directly control many of the key attributes:

- Material types, density, and so on

- Fastening—for example, the number and types of fasteners

- Architecture—for example, modularity, accessibility, and so forth

Engineers have been incorporating recyclable materials in many new-product designs for years. More recently a connection has been made between designing for disassembly and designing for recyclability. In both cases, the goal is to ensure that products are designed in a way that is as attractive as possible to recyclers. Making products quick and easy to disassemble helps.

Clearly, the specific issues and considerations for any DFR methodology will vary by product type, but we can suggest three starting points for understanding the key concepts and regulations pertaining to DFR.

- VDI 2243 is an effort from the German Engineering Society (VDI) to standardize notions about recycling. The purpose of VDI 2243 is to provide engineers with a quick and relatively complete overview of issues to be considered in designing products for recyclability. The guideline provides an introduction to recycling and discussions about production waste recycling, product recycling during a product's useful life, material and waste recycling after a product's useful life, and the application of DFR rules. It contains a wealth of information and illustrates the state of the art in design for recycling in Germany. The Web site www.vdi.de/ provides information about VDI 2243 in German; English and other translations are available through www.beuth.de/.

- You can find an excellent overview of issues to consider in designing for recyclability, prepared by the Georgia Institute of Technology Systems Realization Laboratory, at www.srl.gatech.edu/education/ME4171/DFR-Intro.ppt.

- "2007 Electronics Recycling: A Guide to International Regulations," by Kim Leslie,[4] provides a summary of regulatory developments in recycling and take-back worldwide, covering more than two dozen countries.

A recent example of how engineers can design for recyclability and create both ecological and economical benefits is provided by Subaru's assembly plant in Lafayette, Indiana. The plant was scheduled to produce 180,000 cars in 2008, and the automaker has pledged that virtually none of the waste generated from its output will wind up in a landfill.[5] This places significant focus on the supply chain, where most of the necessary improvements must be made. According to an article in *USA Today* (February 19, 2008), "Copper-laden slag left over from welding is collected and shipped to Spain for recycling. Styrofoam forms encasing delicate engine parts are returned to Japan for the next round of deliveries. Evan small protective plastic caps are collected in bins to be melted down to make something else. All told, Subaru says 99.8% of the plant's refuse is recycled or reused." The reuse effort on the Styrofoam inserts alone has saved the company $1.3 million per year, according to the article.

Take-Back

Producers have traditionally been responsible for the environmental impact of their production facilities, and they have borne the costs of pollution prevention (of course, they passed these costs on to customers). Downstream environmental impacts, however, have often been ignored. As a result, today the concept of EPR has emerged, making producers responsible for environmental impacts over the entire product lifecycle. That's the genesis of take-back policies, in which companies are required to collect and recycle products at the end of their useful life.

At this point, you can safely assume that most products will be subject to take-back legislation in most of the world within the next three to five years. Take-back legislation will provide numerical targets for collection, product recovery, and incineration, along with time frames for implementation. So, if you're starting a product design today, it's important to include this in your overall plan.

In many cases, a third-party company will handle the logistics of your company's take-back program, but make sure you capture the lessons learned so that you can apply them to the design of future products and services.

The initial targets for take-back legislation have been products that create a serious disposal problem in terms of volume or hazardous and toxic content; products for which there are no functioning or active secondary markets; and hazardous products for which the producer does not retain ownership through leasing contracts or other arrangements.

Already, considerable legislation on take-back is in effect or in the works. The European Union and Japan were among the first to introduce such legislation. In the United States, the first take-back law was passed in Maine in 2004. The WEEE Directive, effective in July 2006, makes manufacturers responsible for e-waste, based on recovery, recycling, and collection targets. The legislation is designed with two aims in mind: to create an economic incentive for manufacturers to design more environmentally friendly products, and to reduce the environmental impact of waste by increasing the volume that is recovered and recycled.

The goals for us as engineers will be to find new ways to take back our products with minimal customer costs; increase the "come-back" percentage; implement detailed data collection and reporting (e.g., percentage of products/materials ending up in the waste stream); and measure ongoing improvement in take-back processes.

Chemical and material choices can have both positive and negative business implications for product take-back.

- Substituting for potentially hazardous chemicals from a product can create market differentiation. One of the latest examples is the emergence of household paint products that avoid the use of volatile organic compounds (VOCs).

- Similarly, having hazardous chemicals discovered in your product may damage sales and brand credibility, as we've seen recently with a number of toys.

- Meeting local reporting and disposal regulations can be quite expensive. Avoiding materials that trigger these activities can result in potentially significant savings.

Markets for used materials continue to mature, and some materials, such as aluminum, can have significant value if they are clean. Don't assume that product take-back and processing represent only a cost; they may have the potential to produce income as well.

9

Water and Other Natural Resources

In this chapter, we take a closer look at how the use of water and other natural resources impacts product design. Let's start with the simple, undeniable fact that we're running out of lots of natural resources—both renewable and nonrenewable. A nonrenewable natural resource is one that cannot be remade, regrown, or regenerated on a scale comparative to its consumption. Often, fossil fuels, such as coal, petroleum, and natural gas, are considered nonrenewable resources, as they do not naturally re-form at a rate that makes the way we use them sustainable.

A renewable resource differs in that it may be used but not used up. Fresh water is one example. Other examples include natural resources such as timber, which regrows naturally and can, in theory, be harvested sustainably at a constant rate without depleting the existing resource pool, and resources such as metals, which, although they are not replenished, are not destroyed when used and can be recycled.

Social Considerations

From a societal perspective, every time we use a natural resource we're shrinking the pool for everyone. The immediate and obvious result for consumers is higher prices, but the impact goes far deeper. Beyond depleting the common supply, the process of obtaining these resources from their natural setting can damage the environment. Anyone who's seen the results of a strip-mining operation or clear-cutting in forests can attest to that. The costs of finding more of the scarce resource can be staggering—just ask any oil company executive. Less obvious but equally important is the fact that depletion of natural resources eventually draws the attention of regulatory agencies, and companies are increasingly required to report on their use of these resources (see the "water footprint" discussion later in this chapter). In other cases, these resources come from countries or regions that may not fully comply with accepted employment standards—unfair wages, unsafe working conditions, underage workers, and so on.

Business Considerations

It doesn't take an economics degree to understand that nonrenewable resources will get more expensive as the available supply decreases. And since new products are being created all the time, there's no guarantee that a resource that seems abundant today will remain so in the future. Furthermore, civil unrest and wars can threaten supply.

Since products are often in design for years before going into production, understanding the potential impacts of price changes for your raw materials is vital. It's useful to look at historical price trends, but also to run through some scenarios where a market shock sends prices above their historical trends. This may increase the urgency with which you pursue alternative approaches.

Calculating the Water Footprint

How many liters of water does it take to make a cup of coffee? Sounds like the setup for a joke. Actually, it's an interesting way to envision the **water footprint** of products. In fact, according to Waterfootprint.org, it takes 140 liters of water to make a cup of coffee; it takes 900 liters of water to produce a kilogram of corn; and it takes 16,000 liters of water to produce a kilogram of beef. You get the idea. Studying the water footprint over the lifecycle of a product or service is the only way to get a clear picture of how much total water is embodied in each product or service we design.

The water footprint calculation is fairly straightforward, but with a couple of twists that you won't encounter in calculating embodied energy or greenhouse gas (GHG) emissions.

The water footprint of a product includes the total volume of fresh water used directly or indirectly over the product's lifecycle, and consists of three components: the water use in the producer's supply chain (indirect water use), the direct water use by the producer (for producing/manufacturing or for supporting activities), and the water use inherently associated with the consumption of the producer's products by others.

The lifecycle water footprint further breaks down into three categories: the blue, green, and gray water footprints.

- The **blue** water footprint is the volume of fresh water that evaporated from the global blue water resources (surface water and ground water) to produce the goods and services.

- The **green** water footprint is the volume of water evaporated from the global green water resources (rainwater stored in the soil as soil moisture).

- The **gray** water footprint is the volume of polluted water that associates with the production of all goods and services for the individual or community.

It is relevant to know the ratio of green to blue water use, because the impacts on the hydrological cycle are different. Evaporated water and polluted water are both "lost" (unavailable for other uses).

Trading Virtual Water

Since almost all products and services contain embodied water, when goods and services are exchanged, so is virtual water. For example, when a country imports a ton of wheat instead of producing it domestically, it is saving about 1,300 cubic meters of real indigenous water. The water that is "saved" can be used by the importing country for other purposes. That is why water-scarce countries such as Israel discourage the export of oranges (relatively heavy water guzzlers)—because that in effect prevents large quantities of water from being exported to different parts of the world.

In recent years, the concept of virtual water trade has gained weight in the scientific community and has become the subject of political debate. Trade in real water between nations or companies remains impractical because of the distances and associated costs, but trade in water-intensive products is feasible.

The notion is still is its embryonic stages, but as an analytical concept virtual water trade provides a way to analyze the impacts of consumption patterns on water use.[1] As a politically induced strategy, the questions are whether virtual water trade can be implemented in a sustainable way, whether the implementation can be managed in a social, economical, and ecological fashion, and for what countries the concept offers a meaningful alternative.

Other Natural Resources

There is now clear scientific evidence that humanity is living unsustainably by consuming the Earth's limited natural resources more rapidly than they are being replaced by nature.[2] Many excellent books and articles discuss natural resource usage and sustainability; we recommend that you visit the Appropedia Wiki at www.appropedia.org for an overview of key considerations and recent engineering community insights.

One important consideration pertaining to natural resource usage is often overlooked, however: hidden impacts or unintended consequences. An article posted on Yahoo.com[3] underscores the problem, explaining how coltan, a rare, unrefined metallic ore used in the manufacture of video game consoles, may have fueled violence in the Democratic Republic of Congo:

> *Allegedly, the demand for coltan prompted Rwandan military groups and western mining companies to plunder hundreds of millions of dollars worth of the rare metal, often by forcing prisoners-of-war and even children to work in the country's coltan mines. "Kids in Congo were being sent down mines to die so that kids in Europe and America could kill imaginary aliens in their living rooms," said Ex-British Parliament Member Oona King.*

It's not always possible to foresee every consequence and impact of the natural resources used in the products we design and manufacture, and worldwide supply-and-demand conditions for specific resources are not within an engineer's control. But to the extent possible, think through the short-term and long-term implications of the material selections you make. And if you have concerns, speak up. An ounce of prevention is still worth a pound of cure.

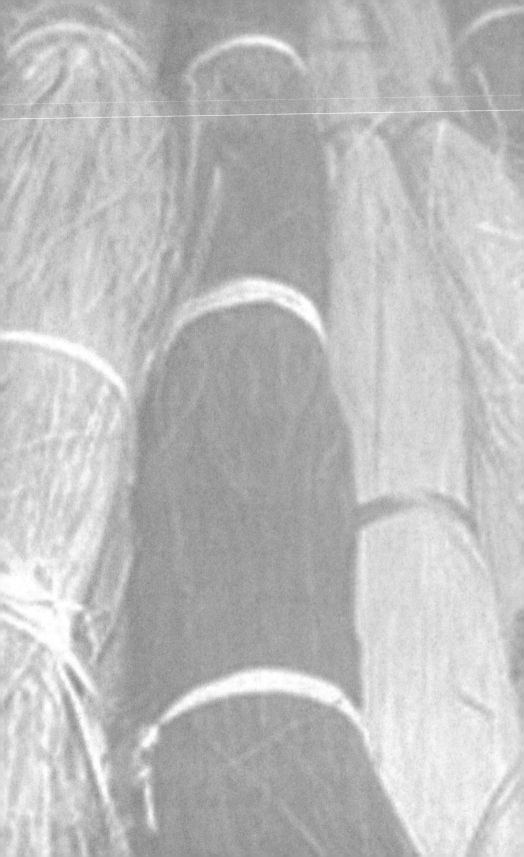

10

An Example of Eco-Engineering: Interface, Inc.

In this chapter, we will look at a shining example of how the lifecycle model we describe, combined with clear priorities, can deliver ecological and economic benefits to real-world companies. Interface, Inc., one of the world's largest carpet manufacturers, has generated outstanding results by leading in "sustainability engineering," which it applies to virtually every facet of its business operations.

Interface is by no means the only example we could cite here, but we find it instructive that the maker of something as seemingly mundane as carpeting—and a leader in an industry long associated with pollution and harmful waste—can set a new standard in eco-effectiveness.

An Aggressive Initiative with Very Specific Goals

In 1994, Interface launched an aggressive environmental sustainability initiative that measured and monitored not only its own environmental impact, but also that of its suppliers. The initiative encompassed environmental sustainability as well as social and economic sustainability. The company also set very specific objectives for its sustainability efforts, laying out business and environmental goals on seven "fronts":[1]

- **Eliminate waste:** Eliminating all forms of waste in every area of business

- **Benign emissions:** Eliminating toxic substances from products, vehicles, and facilities

- **Renewable energy:** Operating facilities with renewable energy sources—solar, wind, landfill gas, biomass, geothermal, tidal, and low-impact/small-scale hydroelectric or non-petroleum-based hydrogen

- **Closing the loop:** Redesigning processes and products to close the technical loop using recovered and bio-based materials

- **Resource-efficient transportation:** Transporting people and products efficiently to reduce waste and emissions

- **Sensitizing stakeholders:** Creating a culture that integrates sustainability principles and improves people's lives and livelihoods

- **Redesigning commerce:** Creating a new business model that demonstrates and supports the value of sustainability-based commerce

Interface uses a lifecycle assessment (LCA) model to calculate global warming impacts from its products. The model analyzes raw material acquisition, product manufacture and transport, and how customers use the products. The CO_2 impacts are also considered based on material types, energy, packaging, and disposal.[2]

At the core of Interface's sustainability efforts is a measurement system that enables the company to understand its impact and change its behavior. These metrics are very specific and detailed. For example, the following are measured and monitored.

- **Net greenhouse gas (GHG) emissions:** Interface calculates its net GHG emissions in accordance with the WRI/WBCSD Greenhouse Gas Protocol. Eighty-one percent of the company's total emissions occur in North America, with only 13% occurring in Europe, due in part to the purchase of green electricity at most manufacturing facilities in the United Kingdom.

- **Toxic chemical elimination:** Steps to eliminate toxic chemicals from its facilities include replacement of ozone-depleting substances (ODSs) in nearly all facilities, elimination of volatile chlorinated chemicals and SARA 313 chemicals, including those that do not require mandatory reporting, and reduction in the number of suppliers, which gives each facility more accurate and efficient tracking capabilities concerning the types of toxic chemicals that enter the company's facilities.

The results for Interface have been remarkably positive from both an ecological and an economic perspective.[3]

- Cumulative avoided costs from waste elimination activities since 1995 are calculated to be more than $372 million.

- Total waste sent to landfills from carpet manufacturing facilities has decreased by 66% since 1996.

- Interface has reduced the total energy used at carpet manufacturing facilities (per unit of product) by 45% since 1996.

- The company's use of renewable energy increased to 27% in 2007. Three facilities currently purchase 100% of their electricity as green directly from the grid, and three other facilities have made 100% of their electricity green through the purchase of renewable energy credits. Interface also generates a portion of its energy through its three on-site photovoltaic arrays, and uses landfill gas in its LaGrange, Georgia, facility.

- The percentage of recycled and bio-based materials used to manufacture Interface products worldwide has increased from 0.5% in 1996 to 25% in 2007.

- On an absolute basis, Interface reduced its GHG emissions by 33% from its 1996 baseline through improved efficiencies, process changes, and direct renewable energy purchases. Interface has further offset its GHG emissions by another 49% through credits from its LaGrange landfill gas project, resulting in a net absolute GHG reduction of 82% (see Figure 10-1).

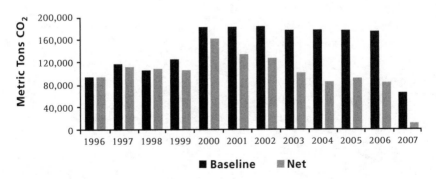

FIGURE 10-1\   Net GHG Emissions at Interface, Inc.

- Water intake per production unit is down 75% in modular carpet facilities, and down 45% in broadloom facilities from 1996 due to conservation efforts and process changes such as eliminating the printing processes at some locations.

- The Interface ReEntry program diverted 133 million pounds of material from landfills between 1995 and 2007.

11

Eco-Engineering: The Grass Is Always Greener

Yes, the title of this chapter has a dual meaning. Among senior executives at global corporations, eco-engineering is already seen as a strategic imperative—even though the practice of eco-engineering is not yet clearly understood by engineers. At the same time, many companies feel constant pressure to prove that their "green" initiatives are greener than their competitors', leading to an upward spiral in greenwashing.

The net result is confusion. It can be difficult for engineers, executives, and consumers to distinguish between an environmentally responsible project and a plain old-fashioned PR grab. If you're an engineer who's really interested in making a positive environmental impact, what should you focus on? Here are a few suggestions, based on lessons learned from the real world of eco-engineering.

Carbon Neutrality: Good Start but Not Enough

Corporate sustainability leaders tend to be a collaborative group. They are open to sharing ideas, swapping stories, and growing their networks of colleagues in other companies. And that's a great thing, because those of us who deal with corporate sustainability on a daily basis know we're in uncharted territory and that we're all learning as we go. We understand it's in everyone's best interest if the overall economy becomes more sustainable. After all, when it comes to climate changes there aren't winners and losers—ultimately we'll either all win or all lose.

For example, let's say your company magically reduces the environmental impact of its operations to nothing so that you're able to deliver your products and services with no impact of any kind. But in the excitement your company decides that you have created such a big advantage through your eco-effectiveness that you better keep it a secret and not share your magic with anyone else. In this case, how much better off is the world? Does erasing the impact of one company make a big difference? Unfortunately, no.

Which brings us to the subject of *carbon neutrality,* the term often used to describe the goal of corporate efforts to lessen companies' impact on the environment. No company can reduce its greenhouse gas (GHG) emissions to zero, so the idea is that Organization A pays Organization B to plant trees, increase energy efficiency, create green energy, or do something else with a positive impact on GHG emissions, thus offsetting Organization A's own carbon emissions.

Many companies have centered their environmental strategy on a goal of achieving carbon neutrality. They are generally doing some efficiency projects, purchasing some green energy, and offsetting the rest. But we've been looking at product and service lifecycles, and we know the part that's within your four walls may be only a small part of your overall impact. What about your supply chain? What about your products in use at your customers' facilities? What about your products at the end of their useful lives?

So, carbon neutrality misses the point. It's good for companies to invest in others' good deeds, but right now it is absolutely critical that companies invest in creating more sustainable versions of themselves.

Yes, we absolutely need a dramatic increase in zero and low-carbon energy sources, but we use so much energy that that won't be enough. We need improvements in basic efficiency—but that's going to be good for only a 20% to 40% improvement, and it won't be enough either. If we want to make serious headway and still have the quality of life we strive for, we're going to need a lot more, and part of that "more" is going to have to be some radical innovation and improvement in the design, manufacture, and delivery of all the products, services, and infrastructure that make up our economy.

More than ever, we need the innovation that comes from competition and open markets. We need companies that view climate change not as a threat but as an opportunity—and are pursuing it with the enthusiasm that big opportunities engender. We need companies to go beyond carbon neutrality to something we, the authors, call "carbon advantage."

You can create a carbon advantage for your company in two ways. First, you can use efficiency and resource reduction to provide a fundamental cost advantage in your operations and products. Second, you can use innovation in green products and services to offer customers a competitive advantage, thus differentiating your offerings.

The good news is there's lots of advantage to be had. Companies that have created more eco-friendly goods—such as carmaker Toyota Motor Corporation and carpet maker Interface—are increasing their market share and improving their business performance.

But more important, there's increasing evidence that we're on the verge of a new, virtuous business cycle: Companies seeking sustainability are looking for sustainable products and services, which provides further opportunities for sustainable companies. As a result, products and services that can help customers improve their own sustainability will be in increasing demand, creating the opportunity for a major shift in market share and a net reduction in business impact on the environment.

Keep your eyes open and you'll see more examples of companies achieving competitive advantage through more sustainable products, services, and operations. Consider General Electric, which noted in May 2007 that revenue for its portfolio of eco-focused products surged past $12 billion in 2006, while the value of its order backlog rose to $50 billion. Chief Executive Jeff Immelt has said the company's Ecomagination campaign has turned into a sales initiative unlike anything he's seen in his 25 years at the company.

Wal-Mart Stores sees the possibility of competitive advantage through increased efficiency and lower cost by reducing product packaging throughout its supply chain. Boeing knows the efficiency advantage of its new 787 Dreamliner can bring savings today, and the value of that efficiency will only grow as awareness of environmental impact grows. At Sun, we've seen the eco advantage in sales of our energy-efficient servers, and as our customers ramp up green IT procurement, this advantage should grow.

In contrast, if Toyota had directed its environmental strategies solely on carbon neutrality, it is unlikely the company ever would have built the Prius. The same goes for the Sun, Wal-Mart, and GE examples.

So, let's not get too hung up on carbon neutrality when carbon advantage can take us so much further. Carbon neutrality is a step in the right direction, but for many companies it's only a very small part of the overall impact they could have. It's in the best interest of those companies, as well as our collective best interest, that they take a broad view and prioritize appropriately across all of their potential environmental opportunities.

At the end of the day, when companies compete on sustainability, the planet will be the big winner.

Greenwashing and Green Noise

It's getting harder for engineers—and consumers—to distinguish between genuine eco initiatives and green hype, and that is a serious threat to eco-engineering. As an engineer, it may prevent your idea from being taken seriously within the company or being understood by your customers. More importantly, if your company is unclear or overly aggressive in marketing the eco benefits of your products, there may even be a negative backlash.

Greenwashing is what corporations do to look more environmentally friendly than they actually are. And lots of corporations are doing it. One recent study examined 1,018 consumer products bearing 1,753 environmental claims, and found that of the products scrutinized, all but one made claims that were demonstrably false or that risked misleading intended audiences.[1] In fact, greenwashing has become so prevalent that CorpWatch, a U.S.-based watchdog organization, now presents Greenwash Awards to "corporations that put more money, time and energy into slick PR campaigns aimed at promoting their eco-friendly images, than they do to actually protecting the environment."[2]

Green noise is related but slightly different. It refers to the information overload about eco-related products and services from corporations, columnists, researchers, market analysts, reporters, and even friends and relatives. The result of too much information is that everything turns to static. *Biodiesel, fuel cells, polyvinyl chloride, sustainability, footprint, all-natural, biodegradable, organic*—eventually the words turn to mush and become meaningless.

What can an engineer do to protect against greenwashing and keep consumers interested in making more environmentally responsible purchase decisions? Here are a couple of suggestions.

Measure and Label

Real data about environmental impact can help fight both greenwashing and green noise. A case in point is the U.S. Environmental Protection Agency (EPA) ratings on new cars. Consider it from the consumer's perspective: You're aware of our climate changes, you're not super-rich, the price of gas has just spiked to another record high, and you're shopping for a new car. It's a pretty safe bet that you're taking a look at the stickers on the car windows that tell you how many miles per gallon you'll get. In the United States, you'll see two numbers: a "city" number and a "highway" number. Obviously, you don't drive on only one or the other of these, but you have a good sense of your driving patterns, so you know how to weigh these numbers in your decision. And you feel good about the fact that a more efficient car is a double win: It lowers your environmental impact and saves you some cash in the process.

The EPA rating is a simple and obvious example—and you'd expect that similar energy-consumption labeling would be available for most consumer electronics products. You'd be wrong. Let's say a new computer is next on your shopping list. Maybe you're buying a new PC for home or a new server for your company. You see that there are lots of eco-rating schemes: Energy Star, 80 Plus, EPEAT, and Climate Savers Computing. With all these rating systems, it must be easy to find out how much power a new PC or laptop will actually use, right? Turns out the answer is no for PCs, and as you go through your day you'll find many other examples where the energy use is not readily available.

The lack of hard data about power makes it harder to make sound purchase decisions; it frustrates people who want to be environmentally responsible; and it may even cause them to make poor decisions in related areas. For example, earlier in the book we talked about the need to cool servers, storage, and network gear in the data center. Without accurate energy numbers, the only thing a cooling engineer can do is oversize the cooling capacity, thus knowingly selecting a less-efficient design.

We've heard a couple of arguments about why this data isn't available. The first is that power varies by application and utilization. That's true for cars as well, and the answer there is to provide more than one data point and let customers extrapolate their own experience. The second is that there are lots of configuration options, and the power varies depending on the exact set the customer chooses. Again, we think customers can deal with the data—let's give it to them.

One approach to dealing with the complexity of some products is to use online power calculators, such as the one shown in Figure 11–1. It's not as simple as a sticker on a car window, but then again, these products aren't as simple as an already built and configured car. An online power calculator lets customers play with different configurations and understand the range of energy impacts that result from their decisions.

With consumers facing numerous energy challenges, they deserve to have an accurate estimate of how much energy the products they buy will use. As engineers, we have the power to give them that information. As we're designing the next generation of power-consuming products, let's take the time and expend the effort to provide hard data about power consumption, and to support processes that standardize the way that data is presented to consumers. Only when we provide open, accurate data on all of the environmental

Netra X4450 Server Power Calculator		
Item	**Quantity**	**Notes**
Processor	E7338 (2.40 GHz 80W) ⇕	Select Processor Type
Number of CPUs	2 CPUs ⇕	Select Number of CPUs
Memory (4 GB DIMM)	None ⇕	Select number of DIMMS (Min 4, Max 32 Total per System)
Memory (2 GB DIMM)	None ⇕	Select number of DIMMS (Min 4, Max 32 Total per System)
Power (AC/DC)	AC ⇕	Select Power (AC/DC)
DVD-R/W Drive	None ⇕	Select number of Optical Drives (Min 0, Max 1 Total per System)
Hard Disk Drives	1 HDD ⇕	Select number of Hard Drives (Min 0, Max 12 Total per System)
PCI-E Card	1 PCIe Card ⇕	Select number of PCI-E Cards (Min 0, Max 8) Select 1 PCIe card for SAS HBA if using internal HDDs. An Internal SAS HBA reduces the maximum to 7 PCI-E Cards
PCI-X Card	None ⇕	Select number of PCI-X Cards (Min 0, Max 2)
Indicate Workload (%):	100	Select Workload range (0% - 100%)

Calculate   Reset

FIGURE 11-1   A Power Calculator for the Netra X4450 Server

aspects of products will we get the kind of green procurement and consumer behavior that we really need.

Read "The Six Sins of Greenwashing"

We suggest that all engineers take the time to read "Two Steps Forward: The Six Sins of Greenwashing," by Joel Makower.[3] It examines environmental claims made in North American consumer markets and finds, as we mentioned earlier, that the vast majority are deceptive, misleading, or simply untrue. The report then describes six patterns in greenwashing and provides recommendations for consumers as well as for marketers.

The key takeaway for engineers is that it's time to get more actively involved in the way the products you design are marketed. The time-honored tradition of throwing finished products over the fence to marketing and PR must come to a close. Insist on reviewing the promotional materials and press releases that are being prepared for your products and make sure all claims are accurate and truthful. Total honesty doesn't have to conflict with promotional goals; in fact, a lack of honesty usually backfires in the marketplace.

One idea is to have a technical eco lead for each product. This person would be in charge of the official eco scorecard for the product and would be the source of true data for marketing. Ideally, this person would also be required to approve all public eco claims before they go out through the marketing department. This may sound far-fetched for your organization, but we've seen examples of it working.

Measuring and Sharing with OpenEco

One of the key themes of this chapter has been the importance of communities, sharing, and transparency to the success of eco efforts and our overall progress as a society. Within that context we'd like to highlight one of our own projects that touches on all of these themes, as well as some of the key community and intellectual property (IP) issues discussed in this book.

OpenEco.org is an online community and Web site sponsored by Sun. We built OpenEco to help organizations measure and lower their GHG emissions, and to make it easy for them to share the data with others. You can calculate direct (Scope 1) GHG emissions from natural gas combustion and indirect (Scope 2) GHG emissions resulting from the use of purchased electricity. The calculation methodology is consistent with the World Resources Institute (WRI) Greenhouse Gas Protocol; the site also leverages resources such as those provided by the EPA Climate Leaders Program to help organizations assess performance measures using common standards.

Figure 11–2 shows a simple formula, which is consistent with the WRI GHG Protocol, for calculating GHG emissions.

$$\boxed{\text{Energy Used (kWh/Therms)}} \times \boxed{\text{Regional Emission Rate}} = \boxed{\text{GHG Emissions}}$$

FIGURE 11–2   Formula for Calculating GHG Emissions

"Originally, we built the core of the measurement tool to help measure and track our own GHG emissions, but we soon realized that everyone else must be facing the same challenges we were," says Lori Duvall, who was instrumental in establishing OpenEco. "We felt a standard methodology was needed because most GHG analysis today is done with spreadsheets or using proprietary tools and can require significant investments in consulting services. We wanted something simpler, less expensive, and more accessible."

All too often, according to Duvall, proprietary GHG analysis tools require engineers to understand many accounting practices and be able to identify the right conversion factors to use in their calculations. Not only does this discourage many organizations from assessing their GHG impacts, but it also means that results are rarely shared outside an individual organization.

"What we recognized was that the scavenger hunt of finding the underlying data is specific to each company, but the calculations and the tables of common data they use aren't," says Duvall. "The data management that you need to do to update each month and track your progress against your goals shouldn't be a unique challenge for every organization either. Gil Friend from Natural Logic helped us generalize what we'd done in a way that would scale and apply to others, and we created OpenEco.org to give everyone a much easier way to get going."

The site also provides a way to compare your GHG emissions with those of other companies in your area, or to determine your GHG per office employee, since it normalizes all of the data and has built-in tools for making comparisons. Doing this kind of comparison is nearly impossible today outside of OpenEco, and is an important step in helping organizations understand where to focus their attention. Use of the site is free, with one caveat: You have to be willing to let others compare against your data. You don't have to attach your name to the data (it can be anonymous), but the data has to be visible for others to use.

"OpenEco isn't a complete GHG tool yet," says Duvall, "but it covers buildings of all kinds, and other sources will be added quickly by Sun and the community. We think it's a great example of how corporations can help harness the power of communities for a common cause—and deliver a useful tool that promotes openness and transparency in a way that's completely consistent with business objectives."

We strongly believe that business and social responsibilities are closely aligned. We're proud of the progress our own company has made in the area of "transparency," from both a business and an environmental perspective, and we intend to continue to pull back the curtain so that everyone can see what we're doing to promote a sustainable and responsible approach that creates value for the ecosystem and our stakeholders. And, oh yeah, the code is freely available as well.

While OpenEco is targeted at organizations, feel free to add in your house or apartment. It will give you a sense of the tool and the power of the community. Hopefully it will help you understand your environmental impact, but more importantly, maybe it will spark an idea regarding something you know or have built and should be sharing with others.

We have begun to collect data relevant to environmental impact from a variety of sources. For example, we now know that more than 90% of Sun's carbon footprint comes from energy use; and we are committed to openly sharing our energy consumption and related GHG emissions data and encourage other organizations to do the same.

Part II Summary, and What's Next

This part of the book dealt exclusively with environmental responsibility—making products and services that meet societal, corporate, and/or personal goals for eco-effectiveness and sustainability. Those are key considerations for every engineer today, and there is enormous opportunity for engineering innovation in every aspect of eco responsibility and at every phase of the product or service lifecycle.

Yet, taken together, the environmental considerations represent just one aspect of being a true Citizen Engineer. Equally important is the way you innovate and use your ideas, and those of others, to create new products and services. The next part of the book is about the responsible use of ideas—how to share them, how to protect them, how to get others to amplify and propagate them—so that your innovations have maximum impact within your communities and companies, and in the marketplace.

PART III
Intellectual Responsibility

This part of the book is about the intersection of ideas and the law. It addresses the question of how to use ideas responsibly—your own and those of others—while understanding the protections of intellectual property (IP) law. Why is this important to a Citizen Engineer? At a macro level, the economic repercussions of IP law can be enormous. At a more personal level, the opportunity is immense. By mastering the world of IP law—patents, copyrights, trademarks, trade secrets, and so forth—you gain a powerful tool for propagating your ideas and controlling the destiny of your innovations. You make it possible to allow others to amplify your ideas, and to create the types of products that are consistent with your ethical goals and your company's financial objectives. Here are just a few of the questions we'll address in this part of the book.

- What's the difference between a patent, a copyright, and a trade secret?

- What's an open source license, and why are there so many?

- If you're working on an open source project, who owns what you contribute?

- How do you protect your work outside your country?

- What are an engineer's responsibilities in protecting privacy?

- When should you sign a nondisclosure agreement—and when should you require others to sign one?

In this part, we provide discussion and practical advice about each of the key IP protection mechanisms: patents, copyrights, trademarks, trade secrets, nondisclosure agreements, and employment contracts. We look at both **inbound IP,** or how you use the ideas of others, and **outbound IP,** or how others are permitted to use your ideas. We also examine the most common variants of open source software licenses and offer guidance for developers. And we offer a few thoughts about how intellectual responsibility can translate into economic advantages for your company—and more control over your ideas and innovations.

Much of the content in this part is applicable in the United States only; if you're working outside the United States, make sure you familiarize yourself with local permutations of IP law. Equally important to note is that we aren't lawyers, so you shouldn't construe anything here as legal advice. The best legal advice we *can* give you is to get to know some good lawyers and seek their participation and counsel.

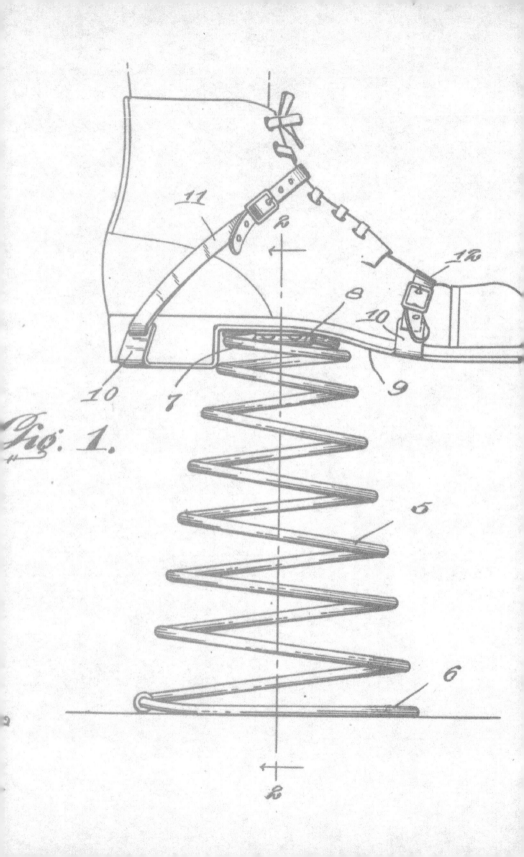

Fig. 1.

12
Intellectual Property Law Fundamentals

> "Good artists copy. Great artists steal."
>
> —*Pablo Picasso*

I t's a world unto itself. It's a field of study for philosophers, legal scholars, and think-tank researchers; it's a battleground for corporations, inventors, and artists; it's a cottage industry for lawyers, consultants, academics, and authors; it's a specialized game with incredibly complex rules and its own arcane language. Collectively, it's called "intellectual property law" and it's dedicated to answering one simple question.

Who owns ideas?

The very notion that someone can own a creation of the mind is one of the strangest ideas conceived by humans—and maybe that's part of the reason intellectual property (IP) laws are so convoluted, complex, and difficult to fathom. But IP laws are a fact of life for engineers, and whether you approve of them or abhor them you must acknowledge that they reflect the underlying value system of our society. As a Citizen Engineer, you're going to have to get to know this strange world. We'll try to make it easier. Do relax a bit—the basic concepts are actually quite simple (but like many aspects of engineering, the details can get complex).

IP 101: Core Concepts

The battle over intellectual property has its roots in a fundamental tension: People who think up new ideas or create new works often want an exclusive right to profit from them—but few ideas are ever completely new and more often build upon the efforts of others.

Intellectual property laws are crafted to protect inventors and creators and the companies that market their works. While it looks like these laws are fundamentally about protecting the "rights" of individuals and corporations, the social value flows more from the theory that a well-engineered system of IP law maximizes overall innovation and progress. If you don't have enough protection, the theory goes, no one can profit enough from an innovation to afford the upfront investment of time and money. With too much protection, you stifle competition and belie that fact that we make the most progress when we build on the ideas of others.

Only four major concepts are involved: patents, copyrights, trademarks, and trade secrets. Here's the super-condensed version:

- *Patents* **protect ideas.** There are different types of patents, including "utility" patents, "design" patents, and "plant" patents. Utility patents are the most common type; if you can meet the required legal standards of utility, novelty, and nonobviousness, you can be awarded a patent that gives you the right (for 20 years from the date of application) to *prevent someone else* from using your idea. In exchange, you have to make public everything that is needed to understand your idea. The biggest misconception is that if you have a patent on something, it gives you the right to make that thing. It doesn't. A patent gives you only the right to exclude others from doing so.

- *Copyrights* **protect the expression of ideas.** For engineers, the "expression" is often in the form of a work product such as software, manuals and specifications, mechanical drawings, chip design, mask sets, and so forth. When you hold the copyright on a work, you have the right to reproduce and disseminate it however you choose. And you get certain legal recourse against others who make copies that you didn't authorize. Copyrights are easier to obtain and last longer than patents, but they protect only the expression of an idea, not the idea itself. Therefore, the protection afforded by copyright is narrower in scope than patents. Others can implement functional equivalents of a copyrighted work without violating copyright law as long as they do not copy the work.

- *Trademarks* **protect the "names" of products that embody your idea.** And by "name" we mean what you call your product, or a symbol or logo you use to identify it and distinguish it from the products of another. If you meet certain standards, trademark law gives you ways to stop someone else from using your name to identify what could be in every respect that other person's identical product. These laws have evolved to prevent confusion among consumers, but they can be very powerful components to things such as open source software business models.

- *Trade secrets* protect ideas that are maintained as a secret. Trade secrets include a broad category of subject matter that may be maintained as a secret, ranging from formulas, manufacturing techniques, designs, specifications, and test methods, to customer, financial, vendor and other forms of information. However, trade secret protection exists only for as long as the idea or information is not generally known or used by others, you derive some economic benefit from it not being generally known or used by others, and you take reasonable measures to preserve its secrecy. The legal protection is against theft, misappropriation, or unauthorized disclosure by others, but not against independent development by others.

While these four areas are, in a legal sense, mostly independent of one another, the interplay among them in IP law can be powerful. A central tool is the **license,** which is a legal grant of defined intellectual property rights by the owner or holder to a third party, typically in consideration for something else. That something else can be as simple as paying money for, say, the right to make copies of a work or to use a patent without risk of prosecution by the holder.

That is, a license is the way you get to set terms on how IP you own can be used. Since you are the owner, you have a lot of freedom in saying what people can and can't do with your IP. The law covering the class of IP (patent, copyright, trademark, trade secret) gives you the protection mechanisms to give your license enforcement teeth.

A very cool interplay of the concepts happens with open source software. We'll go into a lot more detail in Chapter 14, but the basic idea is that a software copyright owner adds (in the comments section of the code) a license. That license, in turn, lays out the terms under which someone else can make copies of the code. Those terms can be as simple as "Copy as you please; just be sure to propagate the copyright notice and this license." Or they can be very restrictive, such as "Not only do you have to propagate the license, but you must republish under the same license terms any improvements or bug fixes and you can't sue people over patents you might have covering this code."

The foundation of open source software is thus in copyright law. If you follow the terms, you can make copies. If you don't follow the terms, you can be prosecuted under copyright law for what would be considered illegal use or reproduction. But first, let's spend some time on the fundamentals and then return to the concept of open source.

The next sections provide basic information about the key IP protection mechanisms, followed by discussion and guidance. Our thanks to Michael Anastasio, law director and associate general counsel with Sun's Intellectual Property Law Group, for his contributions to these sections.

Patents

Patents really aren't what most people think they are. They don't give you the right to actually make anything. But they do give you the right to prevent others from making anything that uses your patented idea. That is, a patent gives you a right to *exclude others*. The simple reason it doesn't directly give you the right to make something is because that thing might also infringe upon someone else's patent, and that other person could exclude you! (And this has given rise to the whole area of patent cross-licensing.) But before we get too far, let's review the basic facts about patents.

- **What does it protect?**
  A patent protects inventions (must have utility—e.g., computer logic, machines, tools, programs).

- **How does it function?**
  It gives the holder a legal right to exclude others from making, using, selling, or offering to sell the invention.

- **How long is it valid?**
  A patent is valid for 20 years from the date of application (in the United States).

- **Where is it valid?**
  Typically, a patent is valid only in the country that granted the patent. A U.S. patent is not enforceable outside the United States.

- **How much does it cost?**
  The absolute lowest you could pay for a single patent is $1,200 U.S. Costs can easily exceed $10,000 when you add legal fees and other associated expenses.

- **What's patentable?**
  Any new and useful process, machine, article of manufacture, or composition of matter, and improvements thereof, can be patented.

Stated another way: You can patent a method, an apparatus, a system, a service, a business method, software, or an improvement to existing technology.

It must be novel: invented by you, not known or used by others, and not described in a patent, patent application, or publication.

It must be nonobvious to one of "ordinary skill in the art."

It must be useful, not simply an abstract idea.

- What's not patentable?
  You cannot patent laws of nature; physical phenomena; mathematics; or literary, musical, dramatic, and artistic works.

- Who owns a patent?
  Patents are granted only in the name or names of the actual inventors. An inventor may sell, will, transfer, lease, or give all or any percentage of the rights to a patent to anyone. This is called **patent assignment**. Patents can also be licensed exclusively or nonexclusively.

A Closer Look: Why Get a Patent (and Why Not)?

There are "defensive" reasons, "offensive" reasons, and other reasons to get a patent. Patents provide protection for 20 years, which gives you the right to exclude others from practicing the invention now and well into the future. Patents can be leveraged to generate revenue through licensing, give your company an exclusive or superior market share, or deter others who may be infringing on your patents from suing you for infringement of their patents. Patents can also generate goodwill and good PR—a reputation for innovation and thought leadership, good citizenship through shared innovation, and evidence of personal achievement for inventors.

A good mental model of a patent is that it is a time-limited monopoly on an idea, in exchange for publicly disclosing the idea. The hope is that by granting inventors this exclusive period, it creates sufficient "air cover" or market opportunity to make a profit on the millions or billions of dollars that were sunk into developing the idea in the first place. In exchange, the inventor must reveal the "secret sauce" in the form of an invention disclosure so that everyone else can see and learn from how it is done. The Latin root *patens* means "open."

Patents can be purchased, sold, leased, or mortgaged, just like any other asset or piece of physical property. Businesses can even donate patents to charities in order to receive tax benefits. Start-up companies use patents, often their only collateral, to lure investment from venture capitalists. Midsize businesses swap and barter patents, even with rivals, to build products they could not make on their own. And many companies license patents to bolster their balance sheets. In fact, the right to profit from a breakthrough idea can be so valuable that the contest over the concept can be more decisive than the competition for consumers, as Sony and Toshiba demonstrated in their tug of war over whose next-generation DVD patents would win out, with Sony prevailing.[1]

And why not get a patent? Well, they can be expensive to obtain and maintain (they often cost tens of thousands of dollars). Also, you might have a business model that relies on free-flow technology, open source software again being a primary example.[2] Or you might not want to disclose your invention to the public, and believe you would be more successful by maintaining it as a trade secret, as The Coca-Cola Company has done with the secret formula for its popular soft drink. In addition, there is a growing movement that casts companies that routinely and aggressively enforce patents as anti-competitive or anti-innovative. Examples are accumulating, particularly in software and business processes,* which have led to reputational damage of the patent holder.

If you decide not to patent something, you still need to be proactive about surrounding IP concerns. If you choose to keep the idea a trade secret, you must avoid publication and take steps to protect it as a trade secret (e.g., by having suitable nondisclosure agreements covering conversations about the idea with third parties). Alternatively, if you are not interested in patenting your invention, but are concerned that another party may attempt to obtain or enforce a patent on the same or a similar invention, you might choose to be very open about your idea and publish it. The advantage here is that you put your idea into the public domain as *prior art,* which may effectively prevent someone else from coming up with the same idea, patenting it, and then going after you!

When to Get a Patent

You should consider applying for a patent when you're working on a new product, a project proposal, or a product improvement or update; when you are writing a technical paper that will be accessible to people outside your group or your company; when you're participating in a standards organization; or simply when you have a cool new idea.

---

* One example is the Amazon 1-Click Shopping concept: see
http://cse.stanford.edu/class/cs201/projects-99-00/software-patents/amazon.html.

Ask yourself the following ten questions, and if you answer yes to any of them you should consider applying for a patent. (Note: Many technology companies have an Invention Disclosure process that you'll need to follow to avoid unnecessary legal entanglements with the patent process.)

- Have I developed a new product or method or a new/improved feature for an existing product or method?

- Did one of my ideas get added to the new-feature list for an existing product or method?

- Did my work produce results greater than expected?

- Have I used a known technology or process in an unusual way?

- Is my development a new step in a rapidly changing technology?

- Did I make an improvement to an existing technology or process?

- When I discussed my work with my coworkers, did they express surprise at the results achieved or the approach used?

- Did my work result in something being better/cheaper/faster (e.g., a more efficient algorithm implemented in software)?

- Did I implement a technical standard or open specification in an efficient, clever, or otherwise improved manner?

- Did my software maintenance work result in an interesting bug fix or significant performance improvement?

If you're trying to determine whether your invention is patentable, the current legal framework discourages you from reviewing other patents! The fact that you've read patents for similar technologies could be cited as evidence that your invention is in fact based directly on a previously patented invention and your application could be denied as a result. Moreover, practicing the invention could be construed as "willful infringement" of another party's patent, which could allow that party to claim dramatically increased damages in a patent infringement suit against you.

This is a rather twisted aspect of existing patent law. You would think that the whole idea behind the patent system is that we could all build off each other's work, with appropriate licensing when needed, or create new innovation by "inventing around" someone's patent. Sadly, the legal system punishes those who do. So, the best advice we can give is to work with a patent lawyer to figure out how your idea relates to other patents. Don't construe this to mean that you should ignore other people's work. To the contrary, a good engineer should be quite aware of other ways that people have attempted to solve a problem.

Applying the Standard of Novelty

You'll notice that patents must adhere to a standard of novelty; the trouble is, it's a very subjective test—one that we ask our patent examiners to perform, backed up by judges and juries who are frequently not well versed in technology.

Until a decade or so ago the patent system in the United States didn't even recognize software patents, and there are still parts of the world that don't. For the most part, software is considered to be a creative work, like writing, and it is covered by copyright law, which provides a more certain and clearer form of legal protection for software versus patent law.

Do Patents Stifle Innovation?

It's important to note that software can be protected simultaneously by both patent and copyright law. That being said, there is also considerable debate about whether software patents are anti-innovative. Amazon's 1-Click patent provides an example.[3] It's an idea. Amazon got a patent on it, so it doesn't matter what code you use to implement it. You can't use that idea, because a patent covers all expressions of the idea. Not surprisingly, the 1-Click patent was controversial from the start, and many of the patent's claims have since been reexamined and rejected, but as of the publishing of this book, other claims have remained valid pending final resolution.

When the system does work properly, it can actually foster collective innovation across competitors. Consider the case of Alexander Graham Bell and Thomas Edison. Bell obtained a patent for inventing the telephone, but the sound quality was awful. Edison came up with an improvement, the carbon button microphone, but he couldn't use it to build a better telephone because Bell had the patent on the telephone. By the same token, Bell couldn't build the better telephone because Edison had the patent on the carbon button microphone. So, what they did, essentially, was cross-license their patents to each other.[4]

When you apply for a patent, you must sufficiently describe your invention so that a person ordinarily skilled in the relevant art can make and use it. You disclose how you did it, and you show others how they can do it too. So, once the patent application is published or the patent is issued, the technology described in a patent is not a trade secret anymore. It's public. In exchange for that, you get a limited monopoly—the right to exclude others for 20 years. More than that, you get control of the destiny of your idea.

"Patents and other intellectual property instruments can give engineers more power to ensure that their ideas are used only for certain purposes—not by tobacco companies, for example, or not in a way that you think would be suboptimal for the public good," says Michael Falk, general counsel for the Wisconsin Alumni Research Foundation (WARF). "From an engineer's perspective, it's about control and it's about prestige—getting credit for what you did and staking your claim on the future of your idea. It's also a way for engineers to demonstrate their value-add to companies, to show that engineering can be a profit center as well as a cost center."

Common Mistakes to Avoid

When the system works, patents are good for the inventor—and ultimately good for everyone—because they encourage inventors to publish their ideas. But you need to be careful. Here are a few common mistakes engineers make that impact patentability or create legal issues or other headaches for companies.

- **Publishing externally:** Engineers often write white papers, technical articles, blogs, or product specs that describe various aspects of a new product or technology. These publications can impact the engineer's (or the company's) ability to protect the IP described. A new product or even a new feature could contain several patentable inventions. Public release of the information makes the invention available to the public—and once that happens in the United States a patent application covering that invention must be filed within one year or the novelty requirement for obtaining a patent will not be satisfied. (In other countries, there isn't even a one-year grace period, so publication before filing a patent application bars one from obtaining patent protection in those countries.)

  To prevent the loss of patent rights, either don't publish anything about a new invention before filing for patent protection, or obtain confidentiality/nondisclosure agreements with outside parties before disclosing any information to them (but do the latter with caution— see the upcoming discussion). Also, it's always a good idea to have a patent attorney review any pending external publication if it may contain information about a patentable invention.

- **Offering customers a "test run":** Sales and marketing will occasionally pressure engineers to give potential customers a sneak preview of a new product before the general release date. These events can impact the ability to protect the IP incorporated in the product, depending on whether the actions constitute an "offer for sale." Once an invention has been "on sale" in the United States the clock begins to tick on the novelty factor, and a patent application must be filed within one year to avoid an "on-sale" bar. Whether there is an offer for sale depends on the totality of the circumstances, and can often involve a complex analysis, even when confidentiality agreements are used. To avoid losing the patent rights, you should always use a confidentiality agreement. In addition, you should always try to make sure all necessary patent applications are filed before disclosing the invention to a customer; and if that is not possible, consult with a good patent lawyer before disclosure.

- **Defining rights for jointly developed IP:** Lots of good ideas arise in collaborative engineering projects, but disputes can arise over who invented what, and who has what rights to what. The "default rules" for jointly owned IP are often not what the parties would have negotiated in an agreement among them. For example, in the United States, absent a legally enforceable agreement to the contrary, each inventor is a co-owner of the entire patent, and each co-owner can freely exploit the patent without consulting with or paying the other owners. Foreign jurisdictions vary. To avoid any misunderstandings, surprises, or unmet expectations, you should always develop an agreement to define the respective rights and responsibilities of each party to a joint collaboration effort. There are many approaches that you can take, and many facets and considerations to such an agreement. You can divide ownership and/or license rights along technology lines or fields of use, you can create separate legal entities to hold ownership of the IP, you can (and should) define roles and responsibilities relative to patent maintenance and enforcement—the list goes on and is extensive. In any case, it's important to clearly specify the contributions, rights, and obligations of each joint owner, and the plan for the exploitation of IP rights, including how they'll be enforced, royalty and accounting duties, rights of assignment and sublicense for the invention, as well as improvements and derivative works, and so forth.

These are just a few examples of measures you need to take when dealing with patents. We'll consider the "dark side" of patent law—and the need for reform—in greater detail later in this chapter. But for context, we first need to examine copyrights, and the interesting interplay between patents and copyrights.

PATENT MYTHS

These "myths" are derived from the experience of Sun's intellectual property attorneys in consulting with engineers.

Myth:
Software, open source software, open standards, and business methods cannot be patented.

Reality:
They can be and often are.

Myth:
A patent grants you the right to practice the invention.

Reality:
A patent grants you only the right to exclude others from practicing the invention.

Myth:
A U.S. patent provides worldwide protection.

Reality:
Patent rights are nation-specific. You must file in other countries to obtain patent protection in those countries.

Myth:
Joint owners of a patent can enforce the patent independently, but must share in any licensing revenue.

Reality:
The opposite is actually true under the default rules.

Myth:
A patent application provides patent protection as soon as it is filed.

Reality:
Enforceable patent rights come into effect only upon issuance of a patent.

Copyright

The term *copyright* is self-defining. It is the right to use, make, or distribute copies. If you write a book or article, you have a copyright on that. You have a right to use, make, distribute, or display copies of your work, and you can license or sell that right to others, who may not do so outside the scope of rights you conveyed to them. Here's a quick recap of the basics about copyrights.

- **What does it protect?**
  A copyright protects the expression of an idea (no functionality—e.g., publications, software).

- **How does it function?**
  It gives the holder a legal right to reproduce an original work (make copies, scan, upload/download, etc.), adapt the work by making derivative works (translations, updates, revisions), distribute copies of the work to the public, and display and perform the work.

- **How long is it valid?**
  Copyright duration is the life of the author plus 70 years; for corporate work the duration is 95 years from the first publication or 120 years from the date of creation, whichever expires first. Thereafter, the work reverts to the public domain.

- **Where is it valid?**
  Copyright is internationally recognized and respected under Berne, the Universal Copyright Convention, the World Trade Organization (WTO), Trade-Related Aspects of Intellectual Property Rights (TRIPS), the World Intellectual Property Organization (WIPO), and others.

- **What's copyrightable?**
  Original works of expression fixed in a tangible medium are copyrightable.

  "Expression" is *not* facts, ideas, procedures, processes, systems, methods of operation, concepts, principles, discoveries, short phrases, individual words, or proper names.

  Only minimal originality is required: Selection and ordering of data is not enough to be copyrightable, but collection and assembly of data in an original fashion embodying some level of creativity is copyrightable.

  "Fixed" media include paper, tape, disks, hard drives, RAM, CDs, DVDs, MP3s, and similar media types.

  Copyright is not available for work in the public domain. That means there is IP risk whenever software is claimed to be "donated."

- **How is copyright claimed?**
  Copyright attaches at the time of creation. In the United States, a copyright notice is not required, but affixing a copyright notice is in

the interest of the copyright holder, both for attribution reasons and for making it more difficult for others to copy your work and claim "innocent infringement" in defense. Copyright registration is also not required unless and until you seek to enforce your copyright.

Copyright is really automatic. If you are the original author, your work is protected under copyright law, regardless of whether you affix a copyright notice, though doing so is advisable. Copyright is, however, generally considered a weaker form of protection than a patent, because the protection of the particular expression of an invention or idea is less valuable than protection of all expressions or implementations of a patentable invention.

The history of copyrights has been very interesting. Consider the plight of early cartographers. They would go through all the trouble of making a map and others would simply copy their work. In the early days of the United States, Americans were unabashedly copying British maps. And that was fine with everyone here in the United States—at first. Then as these native industries started to grow up—map makers and writers of great American novels and so on—it was recognized that these works should be protected.

The duration of copyrights has an interesting and somewhat amusing history (or a disturbing one, depending on your point of view), summarized by this story related by Lawrence Lessig:[5]

> [Initially] . . . copyright law granted protection for the limited time of 14 years. . . . Fourteen years, if the author lived, then 28, then in 1831 it went to 42, then in 1909 it went to 56, and then magically, starting in 1962, look—no hands, the term expands. Eleven times in the last 40 years it has been extended for existing works—not just for new works that are going to be created, but existing works. The most recent is the Sonny Bono copyright term extension act. Those of us who love it know it as the Mickey Mouse protection act, which of course [means] every time Mickey is about to pass through the public domain, copyright terms are extended. The meaning of this pattern is absolutely clear to those who pay to produce it. The meaning is: No one can do to the Disney Corporation what Walt Disney did to the Brothers Grimm.

Copyright infringement, like patent infringement, can be difficult to prove. What constitutes copying? There must be substantial similarity such that the work is recognizable as the original work of authorship. Also, under the doctrine of "fair use," portions of copyrighted material can be used in derivative works such as criticism, commentary, news reporting, teaching, scholarship, research, or parody—without liability for infringement, provided a rather subjective four-factor test is satisfied, which includes consideration of the

- Purpose of use

- Nature of copyright work

- Amount and substantiality of the portion of work that is used

- Economic effect

In general, it's advisable to include a copyright notice on your copyrighted work and not to affix a third-party copyright notice to third-party work that lacks a proper notice. Also, it is illegal to remove third-party copyright notices.

Copyright Using Creative Commons Licenses

For content creators who prefer a spectrum of choices in how others use their work, the Creative Commons provides licenses and tools that let them mark their creative work with the freedoms they want it to carry. Creative Commons is a nonprofit organization that offers alternatives covering the spectrum of possibilities between full copyright protection—*all rights reserved*—and no copyright protection—*donation to the public domain.*

Creative Commons was officially launched in 2002 to counter the effects of what the founders considered to be a dominant and increasingly restrictive permission culture. In the words of Lawrence Lessig, director, Safra Center for Ethics, Harvard University, and cofounder, Creative Commons, it is "a culture in which creators get to create only with the permission of the powerful, or of creators from the past." Lessig maintains that modern culture is dominated by traditional content distributors that seek to maintain and strengthen their monopolies on cultural products such as popular music and popular cinema, and that Creative Commons can provide alternatives to these restrictions.

Creative Commons licenses apply to works that are protected by copyright (books, scripts, Web sites, lesson plans, blogs, and any other forms of writings; photographs and other visual images; films, video games, and other visual materials; and musical compositions, sound recordings, and other audio works). The Creative Commons licenses enable copyright holders to grant some or all of their rights to the public while retaining others through a variety of licensing and contract schemes including dedication to the public domain or open content licensing. The intention is to avoid the problems current copyright laws create for the sharing of information.[6]

You don't need to sign anything to get a Creative Commons license—and all such licenses are nonexclusive. This means you can permit the general public to use your work under a Creative Commons license and then enter into a separate and different nonexclusive license with someone else—for example, in exchange for money.

The original set of licenses grants "baseline rights." The details of each of these licenses depend on the version, and comprise a selection of four conditions.

- **Attribution** (by): You let people copy, distribute, display, perform, and remix your copyrighted work, as long as they give you credit the way you request.

- **Noncommercial** (nc): You let people copy, distribute, display, perform, and remix your work for noncommercial purposes only. If they want to use your work for commercial purposes, they must contact you for permission.

- **No Derivative Works** (nd): You let people copy, distribute, display, and perform only verbatim copies of your work—but they cannot make derivative works based on it. If they want to alter, transform, build upon, or remix your work, they must contact you for permission.

- **Share Alike** (sa): You let people create remixes and derivative works based on your creative work, as long as they distribute them only under the same Creative Commons license under which your original work was published. (See also copyleft.)

Note that the "nd" and "sa" clauses are mutually exclusive. Also, valid licenses that lack the Attribution element have been phased out because 98% of licensors requested Attribution, but are still available for viewing on the Creative Commons Web site. There are thus six regularly used licenses:

- Attribution alone (by)

- Attribution + Noncommercial (by-nc)

- Attribution + No Derivs (by-nd)

- Attribution + Share Alike (by-sa)

- Attribution + Noncommercial + No Derivs (by-nc-nd)

- Attribution + Noncommercial + Share Alike (by-nc-sa)

Sampling licenses are also provided, with two options.

- **Sampling Plus:** Parts of the work can be copied and modified for any purpose other than advertising, and the entire work can be copied for noncommercial purposes.

- **Noncommercial Sampling Plus:** The whole work or parts of the work can be copied and modified for noncommercial purposes.

This book is covered by the Creative Commons Attribution-Noncommercial-Share Alike 3.0 license. We (the authors) want you (the reader) to use any or all of this as you see fit noncommercially, as long as you retain attribution. We have retained the commercial rights to support the business model of book publishers; we think publishers take risks in bringing (new, especially) works to market and thus should benefit from their success.

Additional Concepts: Copyleft and FairShare

There are two other notable copyright concepts you should be aware of.[7] **Copyleft** is a play on the word *copyright*, and refers to the practice of a copyright owner or an original work requiring others to make copies of the original work as well as derivatives and modifications to the work available to third parties under the same freedoms as the original work. For this reason, copyleft licenses are also known as **viral** or **reciprocal licenses.** They are forms of licensing that may be used to modify copyrights for works such as computer software, documents, music, and art. A widely used and originating copyleft license is the GNU General Public License (GPL). Similar licenses are available through Creative Commons, called Share Alike.

FairShare is an idea for a voluntary investment-based patronage system to replace patents and copyright while still ensuring that artists are fairly compensated. It was designed by Freenet creator Ian Clarke, Steven Starr, and Rob Kramer in response to allegations that artists would not receive adequate compensation for their work without enforceable copyrights. In the FairShare system, the investor/patrons would provide venture capital. Clarke envisions that 45% of the money invested in a given artist would go directly to that artist, and another 45% would be given to previous investors. The remaining 10% would be kept by the maintainers of each FairShare service company.

In some ways, this model is similar to a pyramid scheme, but Clarke counters that a vital difference is that nobody would be promised a return on her investment. He argues that regardless of any profits, each patron would have the satisfaction of knowing that she supported an artist whose work she appreciates. Also, earlier investors would profit more, thus rewarding them for investing in artists before they became more popular. The early investors would serve a similar role to studios' talent scouts.

Trademarks

A trademark is used to protect the name, word, symbol, or logo you use to identify your product and distinguish it from other products. Trademark laws help you prevent someone else from using your name to identify a similar or identical product. Whether you're building consumer products or open source software, trademarks can be an important weapon in your IP arsenal. Here are answers to key questions engineers have about trademarks.

- How does it function?
  A trademark helps consumers identify the product and distinguish it from products supplied by competitors or other sources.

- How long is it valid?
  A trademark is valid indefinitely.

- Where is it valid?
  Both state law and federal law apply to trademarks, and typically both state and federal registration is advisable for trademark protection, although rights can attach without registration.

- Why register?
  Registration puts people on notice of your claims and gives you grounds in case someone infringes on your mark.

- What can you trademark?
  "Distinctiveness" determines the strength of the trademark. Arbitrary marks such as "Java™" are strong, whereas marks that merely describe the general nature or function of a product, such as "volume server," are typically afforded no or little protection.

We've noted that trademarks protect the name you give to your creation. In software, for example, others may be allowed to use your code but not your name for the code (typically, the binary build). So, let's think more about the power of software distributions and names. This has nothing to do with how the code itself was generated. But there's something really important going on here.

The only company that has the right to call the Windows™ software bundle "Windows" is Microsoft, meaning it gets to describe the sets of code that come together to create that distribution. And that right of Microsoft has little to do with the fact that it owns the copyright to the Windows code base. For example, the only company that has the right to call a Red Hat™ distribution "Red Hat" is Red Hat, even though those binary distributions are built out from open source components. As it turns out, in the computing business, that's really important because application software on top of operating systems are verified against very specific binary distributions.

Technically, CentOS (http://centos.org) is as close to a perfect reproduction of Red Hat's server software distribution as you can imagine. That's possible for the group of CentOS volunteers to do because Red Hat's distribution is in turn based on freely available open source code. An application that runs on Red Hat almost assuredly runs on CentOS. But CentOS can't claim that, or refer to Red Hat in a multitude of ways, precisely because of trademark law.

That is terribly important because most software packages that are certified to run on Red Hat are not certified to run on CentOS. If you were to call on one of these companies for help and explained that you decided to use CentOS rather than Red Hat, the company will likely tell you that your configuration is unsupported.

It's also important to note that a trademarked brand should always be treated in written communications as an adjective, not a noun. For example, don't call it a Kleenex—it's a Kleenex™ tissue. Don't call it a Snickers—it's a Snickers™ candy bar. It's not Solaris—it's the Solaris™ operating system. Sometimes lawyers can get carried away trying to enforce this principle; for example, we were once advised by legal counsel to stop using the term *Java programmers* and instead refer to them as *Java technology enabled programmers*.

But the thing to keep in mind is that your brand—your good name—has real value. So, even though you give others the right to use your code, you can retain the rights to your trademarked name. That means you keep control over what goes into your distribution, and you live or die by how well it works and how well you support it. Because in the end, it's all about trust and perceived value. Brand matters.

Trade Secrets

A **trade secret** is simply confidential information that derives economic value from not being generally known by others. The subject matter ranges from formulas, manufacturing techniques, designs, specifications, and test methods, to customer, financial, vendor, and other forms of information. Here's a quick summary of basic facts engineers need to know about trade secrets.

- **What is it?**
  Generally, a trade secret is information that is not generally known to the relevant portion of the public and that confers some sort of economic benefit on its holder because it is not generally known. The company must also make reasonable efforts to maintain its secrecy.

- **What is protected?**
  A trade secret effectively allows a perpetual monopoly in the secret information.

- **How is it protected?**
  Protection typically is very narrow and limited since anyone can independently develop and use information that may be a trade secret; however, theft, misappropriation, or unauthorized disclosure by a third party gives rise to liability, and if it is done with intent, it can result in criminal liability, including imprisonment.

- **How long is it valid?**
  There is no set term, but also no minimum term guarantee. Once lost, it is lost forever.

- **Where is it valid?**
  A trade secret is valid in the state in which the trade secret is kept (it is defined by state law, not federal law).

Coca-Cola™ provides the best known example of how trademarks work: The formula is not patented but has been a trade secret for decades; it confers clear and obvious economic benefit on The Coca-Cola Company; and the formula is aggressively guarded and allegedly has not yet been discovered by any competitor, even though reverse engineering is allowable. And because of trademark law, even if someone made a cola that tasted exactly like Coke, that person still could not use the name "Coke" in describing his product.

Trade secrets are not protected by law in the same manner as patents or trademarks. In the United States, they arise out of state laws, though there are some federal laws such as the Economic Espionage Act of 1996 that also apply protections to trade secrets. Most states have adopted the Uniform Trade Secrets Act (USTA), and some with some variations, while a few states have their own statutes and continue to apply common law.

One of the most significant differences between patents and trademarks and trade secrets is that a trade secret is protected only when the secret is *not* disclosed. Owners of trade secrets try to keep their special knowledge out of the hands of competitors through a variety of civil and commercial means, including the use of nondisclosure agreements and restrictive employment agreements that include noncompete clauses. Violations of these agreements can result in substantial financial penalties. Similar agreements are often signed by representatives of other companies with whom the trade secret holder is engaged, such as in licensing talks or other business negotiations.

There is no set expiration date on a trade secret. Trade secret protection can, in theory, extend indefinitely, which may offer an advantage over patent protection, which is of limited duration. However, a third party is not prevented from independently creating and using your trade secret information.

As noted earlier, trade secrets are protected from being stolen, disclosed, or otherwise taken without the owner's consent, referred to as **misappropriation**. Misappropriation can result in substantial civil liabilities for the party that engages in misappropriation, and if it is done with intent, it can result in criminal liability, including imprisonment.[8]

Even if your ultimate objective is to seek patent protection for an invention, trade secrets will play an important role. This is so because in order to preserve the novelty of your invention to meet the requirements of patentability you will need to maintain the invention as a trade secret. Nondisclosure agreements are a useful tool in this regard.

Nondisclosure Agreements

The way you can protect trade secrets and other confidential information from unauthorized use or disclosure is with a nondisclosure agreement (NDA). Here's a recap of the basics about NDAs.

- **What does it protect?**
  An NDA protects the confidentiality of secret information disclosed during business-related transactions.

- **How does it function?**
  Parties sign a contract agreeing to maintain the secrecy of specified information; violation can result in large liabilities.

- **How long is it valid?**
  An NDA is valid for whatever period is specified in the contract.

NDAs can be either "mutual" or "one-way." In a mutual NDA, both parties disclose confidential information to the other and the other must agree to maintain the information confidentially and not to disclose or use it for any unauthorized purpose; in a one-way agreement only one party does so. An engineering team, for example, may require a new-hire engineer to sign a one-way NDA prior to disclosing to the new hire information that she will need to know in order to work on the project.

Many companies have standard NDA forms. Generally, these forms cover four key elements: defining what information is considered "confidential information"; defining what information is excluded from the definition of confidential information or from the confidentiality obligations; the obligations of the receiving party with respect to nondisclosure, nonuse, and disposition of confidentiality information; and the time periods of the confidentiality. Whether you're asking others to sign your company's standard NDA form or being asked to sign another company's NDA form, always have a lawyer read it first. Make sure the terms do not effectively waive any claim of trade secret confidentiality, or operate to convey any ownership or license rights to your IP. Your inadvertent waiver could result in the loss of your company's trade secrets or other IP and leave you with no legal recourse.

In the course of meeting with colleagues, participating in standards organizations, or attending presentations about emerging technologies, you may find yourself in a situation where you feel you're hearing confidential information when you haven't signed an NDA. For example, you may decide to sit in on a presentation about a new approach to asynchronous chip design and realize halfway through the presentation that their idea is similar to one you or your colleagues have been working on. What should you do?

First and foremost, you should do everything you can to avoid those situations. You can expose yourself and your company to infringement and/or misappropriation or similar lawsuits down the road even though the other party voluntarily disclosed the confidential information, and if you didn't sign an NDA. In any meeting you attend where you know confidential information will be discussed, you should consult with counsel to get advice on the appropriate precautions to take, and also be sure that an appropriate NDA is in place.

Employment Contracts and IP Ownership

When you're hired as an engineer at most companies today you'll be asked to sign an employment contract. It will spell out your compensation, your benefits package, your required hours of work, the conditions under which you could be terminated, and—somewhere within multiple pages of dense legalese—the company's policies regarding who owns the ideas you come up with on the job. But what about the ideas you came up with before this job? And what about the ideas you may conceive while working on your own time, at home, on an open source project?

Many employment contracts today specify that **the company owns anything its employees conceive or develop** that are *"related to the business of the company"* or *"during the course of employment."* And many contain clauses that stipulate you can't do anything *"to the detriment of the company"* or that is *"in direct competition with the business of the company."*

At first glance, these terms sound straightforward enough. After all, if you're being paid to do engineering work, the company expects you to invent on its behalf—and the company stays in business by exploiting the innovations you come up with to create new competitive advantages. But when it comes to IP ownership in today's engineering environment, the issues are seldom straightforward. Here are some examples.

- **When is "during the course of employment"?** In other words, when is company time and when is personal time? For most engineers there's really no such thing as being "off the clock." You have a company-supplied cell phone that you use to talk to friends and colleagues at all hours; you use your company-supplied PDA or laptop for both work and play; you have social networking accounts that include both professional and personal contacts (in one recent court ruling, a journalist was forced to hand over the contents of his Facebook contact list to his former employer after he left the company[9]). And ideas can pop into your mind at any time. What if you conceive an idea that could be a breakthrough for your company while you're mowing your lawn? What if you're working on an open source project and you have a brilliant idea that could also benefit your company? Or you're sitting in your company-supplied office and you have a great idea that would help an open source project? What if your idea doesn't relate to the company's current business, but could possibly lead to the company getting into a new line of business?

- **What is "detrimental" to the company's business?** What if the idea you have for an open source project winds up creating new competition for your company? What if your participation in an open source project takes away from your productivity at work?

- **When can you participate in an open source project?** Do you need to ask permission—and if permission is granted, who owns your contributions? And what if your company requests that you work on a specific open source project? What rights do you keep and what do you give away?

As you can see, these issues are nettlesome. Here are some general guidelines to help you sort through it all (again, this is not legal advice; always consult an attorney for guidance about your specific situation).

- **Read and understand the terms of your employment contract.** This should go without saying, but it's astonishing how often engineers later find themselves dismayed by the terms they've agreed to. Read the fine print and know exactly what you're signing.

- **Assume "my time" and "company time" are indistinguishable.** The notion of personal time is becoming increasingly difficult to isolate from work time, and today you'd be taking a huge risk if you expected a court to side with a premise that you conceived and developed a novel idea entirely on personal time. Your company may have a clear definition of what constitutes personal time as opposed to company time, but chances are it doesn't matter "when" or "where" creative genius struck; the more relevant issue is whether your idea pertains to the company's business. If it does, the company will probably own the rights to it unless there is some explicit, written agreement to the contrary. If your work doesn't relate to the company's business, beware the possibility that the company may one day get into that business. Often there are **skunkworks projects** or small teams working on specific innovations secretly; don't assume you're privy to everything that's going on in your company.

- **Document your work thoroughly—even work you consider to be on nonrelated activities—to establish exactly what you did and when.** This could be important in the future to establishing intellectual property rights (particularly if you file for a patent). If there's a possibility that your company may get into the "business" of what is currently your "hobby," make sure your previous work can be carved out in your employment contract or specifically identified as predating the company's interest.

- **Consult an IP attorney if you have any questions or concerns.** If your employment contract is unclear about any aspect of IP ownership—whether it concerns participation in open source projects, carve-outs for previous inventions, or other terms and conditions—talk to an attorney. Your local bar association can usually help you find a quali-fied IP law specialist.

Previous Inventions

Many engineers are owners or joint owners of multiple patents, copyrights, or other forms of intellectual property, and accounting for these "previous inventions" can become complicated as engineers change jobs or roles within an organization. Here are some specific suggestions and guidelines pertaining to previous inventions.

- **Take the time to itemize every innovation for which you own a patent, a copyright, a trademark, or any other form of IP owner-ship—before you even interview for a new job.** Typically previous inventions can be "carved out" in your employment contract, or specifically separated from any additional innovations you come up with on the job. Include all intellectual property you co-own with others. Be sure you have full documentation for everything you claim as your IP. This could also become important in establishing the novelty of future inventions; it can be used in comparisons to "prior art."

- **Don't disclose previous inventions that could violate a prior confi-dentiality agreement.** Instead, you may simply disclose a "cursory name" for the invention, a listing of the party(ies) to whom it belongs, and the fact that full disclosure has not been made for that reason.

- **Expect your company to claim a license to your prior invention if you use it in a new product.** Many employment contracts specify that if you incorporate your previous invention into a "product, process, or machine" of the company's, the company will have a "nonexclusive, royalty-free, irrevocable, perpetual, worldwide license" to "make, have made, modify, use, and sell the prior invention," or other words to that effect. This is quite reasonable from the com-pany's perspective; otherwise, it would be at risk of employees pur-posefully including its previous inventions in products, then demanding the company pay a license fee once it is too late. If licens-ing your prior inventions is not acceptable to you, you must ensure that you do not incorporate them into your new company's products; you must negotiate different terms in your employment contract or not sign the contract.

Participation in Open Source Projects

In recent years, participation in open source projects has become increasingly common among software engineers. This has opened new avenues for creative expression and participation in diverse projects, but it has also made life more complicated from an IP ownership perspective. Here are a few IP guidelines for those who wish to contribute to open source projects while an employee of a commercial enterprise.

- **If you want to participate in an open source project, check your IPR clauses first.** Examine the intellectual property rights (IPR) clauses of your employment contract and make sure you are explicitly permitted to participate in an open source project. Then be sure you understand and agree to your company's policies regarding ownership of your contributions to that project. Some employers have set policies regarding open source projects but are open to discussion about specific projects and IP ownership/exploitation rights; others actually encourage or even mandate participation in open source projects.

- **Know the open source community's policies on IP ownership.** Remember that source code for software is copyright; code contributed to open source projects remains the copyright of the original copyright holder; and copyright holders have the right to dispose of their material any way they see fit. Some open source projects, notably Apache and Samba, have policies that prevent individuals—hence the companies that hired those individuals—from retaining copyright ownership. Other projects insist on the assignment of copyright. In some projects, every contributor owns the copyright for his contribution. In any case, only the copyright holder has the power to reassign his copyright or license his copyright material. If you are not the copyright holder, you must obtain the explicit consent of the copyright holder for anything you wish to do with his material.

- **Know your company's policies on IP ownership for open source contributions.** Find out whether your company has a policy that dictates who owns the copyright for open source contributions by engineers. Is it owned by the company, the employee, or both? In most cases, the company will retain the copyright ownership (or acceptable license) so that it can implement dual licensing should the need arise.

- **If you ask permission to work on an open source project, get the response in writing.** If you agree to an IP ownership arrangement with your company that is different from the terms contained in your employment contract, you should make sure both your request and the company's response are thoroughly documented. In one case that made headlines recently, DDB Technologies, a holding company, sued MLB Advanced Media (MLB) for allegedly using its patented technology without a license on many of its online games. What happened, in a nutshell, is that an employee of Schlumberger approached his company's legal department and requested verbal permission to work on computerized replays of sporting events on his own time (implying that the employee would own any IP he created in this technology). He subsequently patented his work, left Schlumberger to create his own company (DDB Technologies), and wanted to license the patented work to MLB. Allegedly, MLB at first showed no interest, but later used the inventions in some of its online games without compensation. The former Schlumberger employee claimed that Schlumberger had granted his request for permission to develop and "retain ownership" of the new technology; however, he did not have adequate documentation of Schlumberger's response, and as such, Schlumberger retained IP ownership. DDB Technologies then sued MLB because the employee's employment agreement automatically assigned the patent rights in question to Schlumberger. In the end, MLB paid a licensing fee to Schlumberger, not to the former employee whose innovations were used.

Sun's acquisition of MySQL AB in December 2007 provides a good example of why IP ownership issues have become so complex in the open source era—and how established enterprises are becoming more accommodating to engineers who wish to participate in open source projects.

MySQL is the world's leading open source database and has always been a leading advocate of the "open source lifestyle," where contribution to open source projects is not only accepted but actively encouraged. At the time of the MySQL acquisition, Sun was also a leading open source advocate, but tended to be very selective about the projects in which its employees were encouraged to participate. Understandably, Sun had concerns about allowing its engineers to contribute to projects such as Linux, where they could potentially create something that was detrimental to Sun's core business, or to other projects where they would not own or control the IP they created on behalf of the project.

Sun's answer: Rather than create a "white list" and a "black list" of open source projects, and rather than require engineers to get explicit permission to participate in an open source project, Sun essentially opened the door and allowed engineers to contribute to any project—on the simple condition that they let Sun know. There is now an easily accessible Web site and a standard disclosure form engineers can use to keep the company posted about their activities in various communities.

The response from Sun engineers has been overwhelmingly positive so far. There is a sense that Sun is actively encouraging participation rather than "allowing" it; engineers have more outlets and options for expressing their creativity; Sun is finding it easier to retain talented engineers; and potential sources of conflict of interest have been reduced.

Tip Sheet: Inbound and Outbound IP

Remember that IP law is there to protect both your use of the ideas of others (inbound IP) and the use of your ideas by others (outbound IP). Table 12-1 is a quick-reference guide to using the key IP protection mechanisms responsibly—whether you're the source of the IP or you're using someone else's IP.

TABLE 12-1   Reference Guide to IP Protection Mechanisms

PATENTS

Inbound:
- Don't do your own patent searches or read other patents to determine whether your invention is patentable, without first seeking the advice of patent counsel.
- Perform a good-faith investigation to avoid inadvertent patent infringement under the direction of patent counsel.
- Avoid making any written or email comments about the patentability or potential infringement of a patent, except under the direction of patent counsel.
- Don't make, use, or offer to sell any invention that you know is covered by a patent held by another party unless you are sure you have an adequate license from the holder.
- Cut the risk of willful infringement by notifying your legal department immediately if someone informs you of a potential patent issue.

Outbound:
- File for patent protection immediately before the invention is made public or disclosed to a third party.
- Avoid the default rules for jointly owned patents.
- Clearly define IP ownership, rights, contributions, and obligations in joint development projects.
- Create specific plans for enforcement, duty of royalty and accounting, and assignment and license of rights for the IP and any improvements/derivative works.
- Avoid previews, test runs, or external presentations involving potentially patentable IP until all patent applications have been filed.

Continues

TABLE 12-1   Reference Guide to IP Protection Mechanisms *(Continued)*

COPYRIGHTS

Inbound:
- Don't assume that a work is not copyrighted because it doesn't carry a copyright notice.
- Never remove a third-party copyright notice.
- Most open source software is copyrighted—make sure you know and comply with the precise terms of the license before you use or incorporate any part of it in your work.
- Before you rely on "fair use" as a defense to infringement or a reason not to seek a license from the copyright holder, consult with an IP attorney, and remember the four-factor test (purpose of use, nature of copyright work, amount and substantiality of work used, and economic effect).

Outbound:
- Always use a proper copyright notice in your copyrighted work.
- Remember that the company owns the copyright, not the individual, absent an agreement to the contrary (development agreement, contribution agreement, etc.).
- Try to avoid co-ownership of copyrights.
- Consult an IP attorney before mixing open source software with different licensing mechanisms.

TRADEMARKS

Inbound:
- You cannot use an existing trademarked product name as part of the name of a new invention or in a manner that could create confusion as to the source of a product.
- Your written materials do not need to acknowledge third-party trademarks.

Outbound:
- You don't need to register a trademark to attach rights, but registering a trademark can strengthen your trademark protection and claims.
- When writing about a trademarked product, use the trademark as an adjective, not a noun; for example, "Solaris operating system," not simply "Solaris."

TRADE SECRETS

Inbound:
- Remember that theft or misappropriation of a trade secret is a violation of the law, and if done with intent, can constitute a federal crime.
- Never use other employers' trade secrets at your company, or use your company's trade secrets at other employers.

Outbound:
- Keep trade secrets physically and digitally secure.
- Always use NDAs in any oral or written discussion of a trade secret.

NDAs

Inbound:
- Avoid participating in nonpublic discussions about an invention, product, or technology that is similar to anything you're working on, unless you have first consulted the advice of counsel and implemented appropriate measures to avoid claims of infringement, misappropriation, or the like.

TABLE 12-1   Reference Guide to IP Protection Mechanisms *(Continued)*

Outbound:	• Always require anyone who will be receiving information about an invention, product, or technology enhancement that is nonpatented or nonpublic to first sign an NDA.

How to Protect Your IP in Emerging Markets

The WTO's Agreement on Trade-Related Aspects of Intellectual Property Rights, or TRIPS,[10] has made strides in harmonizing IP laws worldwide, but the legal regimes protecting IP still vary widely from one country to the next. China, India, Brazil, and other emerging markets still have rather murky IP reputations—places where enforcement of patent, trademark, and copyright law is uneven or minimal and counterfeiting and piracy can run unchecked.

However, there is reason for optimism. Countries with fast-growing consumer classes have a growing interest in protecting their own IP rights, and recent victories in Chinese and Indian civil courts by global companies highlight a commitment to improving IP protection.

- In April 2006, 3M successfully sued a Shanghai-based manufacturer for infringement of 3M's Chinese patents for respirator masks.

- The following June, a Chinese court upheld the validity of Pfizer's Chinese patent for Viagra and issued an injunction against two infringers.

- On May 23, 2007, the Supreme Court of India in a landmark ruling ordered Dabur to stop using a package design that is deceptively similar to Heinz's Glucon-D packaging.

While the slow march toward uniform IP law and enforcement continues, there are steps engineers and companies can take to safeguard their IP in emerging markets. Alicia Beverly, chief IP strategist with IP Wealth, offered the following tips for protecting IP in China, the largest emerging market.[11]

- China is not a DIY (Do It Yourself) country. Get professional help.

- China is a "first to file" country with no recognition given to use or ownership by other parties. It is therefore essential that you file for your rights—trademarks, patents, etc.—*before* you enter China. Failure to do so is an invitation to the manufacturer or distributor you are working with to do it themselves.

- Contracts must be translated into Chinese, cannot be common law-centric (the United States, England, Canada, and Australia are all common law countries), and must cover everything because anything omitted is fair game.

- Investigate whether your current trademark is useful for the Chinese market, conduct searches, and then protect several versions of your trademark—the English version, the Chinese translation, and even a phonetic version of the English version.

- Design products that are harder to imitate and commit to continuous innovation to keep one step ahead.

- Consider splitting elements of your production in different locations.

You can also find general information and guidance for protecting IP in emerging markets in a variety of places, including "The Protecting Product IP Benchmark Report: Safeguarding Design Intellectual Property in a Global Market" by Aberdeen Group.[12] In addition, here are a couple of specific suggestions from David McHardy Reid and Simon MacKinnon, who have considerable business and research experience in China.[13]

- **Be quick with patent and trademark registration.** Foreign companies entering the Chinese market are sometimes surprised to find that patents on key elements of their products or technologies already have been filed by someone else. This is an entirely legal maneuver, as patents are awarded to the first to file, not necessarily to the originator of the product or technology. Usually these patents have been filed by a Chinese company that uses them either to gain an edge against foreign competition or to negotiate a lucrative agreement to transfer the rights to a company that wants to enter the Chinese market. Trademarks present similar problems. Most products sell better in China if they are given a Chinese name. Some companies in China register trademarks that would be suitable translations of the names of foreign products, again either to sell a competitive product or to negotiate the rights with a foreign company. The only solution for foreign companies is to file patents and Chinese-language trademarks in China as soon as possible.

- **Research and keep up with best practices.** Many foreign enterprises fail to adequately search for and keep up with best practices in protecting intellectual property in China. Information is readily available through the many chambers of commerce and trade associations operating in China. One group dedicated to this issue is the Quality Brands Protection Committee, whose members include many major multinational companies.

For more about protecting your IP in China, check out "China's Trademark Laws—Simple and Effective" (www.chinalawblog.com/2006/09/chinas_trademark_laws_simple_a.html), which talks about how Chinese trademarks are indeed good protection, and "Nike on China IP Protection: Just Do It with Green Tea" (www.chinalawblog.com/2006/07/nike_on_china_ip_protection_ju.html). Both contain a number of additional good suggestions for protecting your intellectual property in China.

Back to Patent Protection: The Good, the Bad, and the Ugly

In previous sections of this chapter, we discussed that software can be patented and/or copyrighted. There are some intriguing interplays here that engineers should be aware of. When you take a closer look at current patent laws, the decision whether to patent or to copyright isn't always as clear-cut as it seems.

In the software industry, copyright appears to be (mostly) better for balancing the interests of individual innovators and the general public. Copyright puts the licensing control clearly and explicitly in the hands of the developer. And a copyright notice can include a license for any related patents the developer may have obtained. Some of the most effective of these are "patent peace" grants, such as what Sun has done with the Common Development and Distribution License (CDDL): You get a grant to our patents as long as you follow the copyright license, including the provision that you won't prosecute for any patents that you might have. If you do prosecute, our patent grant is revoked.

So, in a way, the utility of a software patent here is that it can put more teeth into the potential enforcement of a public license. That's fine when used in this way. But any developer is always open to attack from a patent troll.

In such cases, it's rare that the patent holder is actually putting the idea to use himself; he just wants to charge you an exorbitant fee. So, patent peace doesn't come into it. At least we at Sun have some resources to combat this. In 2004, we paid $92 million to Kodak to immunize the entire Java community from infringement claims on mechanisms that Kodak itself doesn't use—the patents were acquired from a third party. This felt like insanity, but we determined that we had to pay in order to indemnify the whole Java community.

A patent is a far blunter instrument than a copyright and tends to teach us far less than the code itself. Developers don't sit around reading patents to understand some new software pattern or idea, and remember, the limited monopoly we grant a patent holder is in exchange for teaching others how to do it so that when the patent expires everyone is better off.

Another hitch with the patent process is that today's patent system has also proven to be easy for opportunists to manipulate. In recent years, there has been a huge influx of new patent applications. The total number of patent applications filed around the world has increased steadily, particularly since 1995, according to WIPO. The United States had more than 900,000 patents pending in 2005. The Japanese Patent Office also had more than 800,000 patents pending in 2005, and the workload at certain patent offices has increased faster than the capacity to examine patent applications.[14]

Part of the growth in patent applications is predictable given the growth in worldwide engineering activity. But part of the growth is also due to new forms of patent system exploitation—people filing for patents on marginal ideas simply because they believe (often rightly so) that the patent will be granted. In effect, they're exploiting the backlog of applications and the increasingly technical nature of many patent applications.

Why would people do this? One reason is an exploit called **patent trolling**. If you've been issued a valid but overly generic patent, it may be possible to use it to secure cash settlements from companies that are seeking to develop products incorporating a similar technology or innovation. Microprocessor designs, for example, can incorporate thousands of individual patents. Someone who holds a patent on a technology that could be construed as similar to a technology being used in a new microprocessor design could claim patent infringement and demand compensation. The result for the microprocessor company could be an unexpected cash outlay, or significant delays in the product development cycle and missed market windows.

So, the challenge is to create a system that promotes innovation, not simply "patent pooling" or accumulation of the deepest stack of patents. If the system is too easily manipulated, or if the potential consequences of making an IP mistake are too great, inventors are disheartened.

"In general, the patent system works pretty well—both in terms of spurring innovation and in protecting inventors," says Michael Falk, general counsel for WARF. "I do think, however, that there are cases where the patent system can cause excessive friction and inhibit innovation. In the area of wireless technology, for example, there may be some companies that weren't funded because there were too many IP issues and potential investors were afraid to move forward. That's a cost of the system—without the friction you'd have more innovation—but more often the benefits outweigh the costs."

We believe the patent system in the United States needs reform, and we're far from alone. In the words of David Evans, senior vice president at NERA Economic Consulting, Inc.:[15]

> The U.S. Patent Office lacks the resources to distinguish inventions that deserve protection from those—the silly, the obvious, and the hardly new—that do not. That problem is exacerbated by the legal system. Once granted, patents are hard to void, and juries and judges tend to favor the patent holder. Uncertainty over the scope and even existence of patents also creates expensive litigation. For instance, holders of "submarine patents" have launched huge royalty claims after having kept a patent secret, quietly waiting until the technology or business process becomes an industry standard. . . . In changing the system, policymakers ought to remember that extremists from both ends of the protection spectrum cannot be trusted. Recommendations to grant blanket elimination of patents in certain industries—such as the claim that there should be no software patents—should be treated as skeptically as demands for blanket extension of patent rights to areas where they do not now exist.

And patent reform is underway, albeit excruciatingly slowly. According to the *New York Times* (January 13, 2008):[16]

> Congress is expected to take a hard look at the nation's patent laws this year in what could result in the biggest change in generations . . . The House passed the Patent Reform Act in 2007, with the Senate expected to pass its own version in a matter of months. A key element in the debate centers on whether to limit damage awards in patent infringement lawsuits—a provision sought by a host of the giant technology companies, including Microsoft, Apple, Intel and IBM. . . . The real question for lawmakers this year is how to rewrite the patent laws to best protect innovation while limiting needless litigation and abuse of the patent system. Stay tuned: the fight has been engaged, and there will be heated arguments about who is right, who is wrong and how technology advancements should be protected at a time of rapid change in the industry.

13

Open Source Software: Licenses and Leverage

"Innovation happens elsewhere. Most of the smart people don't work for you."

—*Bill Joy, cofounder, Sun Microsystems*

A funny thing happened on the way to the twenty-first century: The Internet ended up enabling the free exchange of ideas within communities that span both corporate and national boundaries. Surprising to many, these open communities—especially in software—ended up out-innovating their proprietary counterparts.

Certainly open source software technology *adoption* can ride wild exponentials upward as the technology user community explodes around the typically much smaller nucleus of code contributors. It has also been proven possible to build quite valuable businesses around open source software. Red Hat, for example, has been enormously successful in redistributing the free and open source software components of the Linux kernel and layered applications. MySQL AB, maker of the MySQL open source database, fetched $1 billion from Sun Microsystems in January 2008. Even certain MySQL *binary distributions* are free. (On a typical day more than 50,000 of them are downloaded worldwide!)

The open source model has proven that at times, the best way to control a technology is to give it away. And by "give it away" we mean "release the code under an appropriate open source software license."

In this chapter, we detail how different licenses work and how and why you want to choose one over another. At a basic level, however, all open source licenses work the same. They are just that: software licenses. The license is written in comment lines at the top of a source code file and says, in effect: "You can freely make copies of this code, but only if you follow the rules spelled out here. If you don't follow the rules, then we retract the license and any copy you make will be considered a violation of our copyright." Here are some of the kinds of rules you'll find.

165

- **Propagation:** Virtually every license requires you to propagate both the license and the original copyright notice.

- **Put-back:** You might be required to republish any changes you made, whether they are optimizations, added features, or bug fixes.

- **Virality:** You might be required to ensure that any "adjacent" code modules carry the same software license, for software you wrote yourself or obtained from other sources.[1]

- **Patent grant:** By using the code you implicitly grant any patents you might have on that code that might infringe to all other users of the code.

- **As-is:** There is no warranty that the code actually works, or isn't somehow infringing on someone's patent. And by you using the code, you won't sue for liability.

- **Noncommercial:** You can use the code for personal, experimental, or academic uses, but you can't sell a product or run your business on it.

At one end of the spectrum are licenses such as BSD and MIT, which are very permissive and mostly just specify propagation. At the other end are very prescriptive licenses, such as many forms of the General Public License (GPL) that are viral and require put-back and patent grants (GPLv3), among other things. The Apache and Mozilla licenses, as well as the Common Development and Distribution License (CDDL), are somewhere in the middle. These are all spelled out in this chapter.

A good mindset is to think of an open source license as a tool. You have some objective in mind and you should use the right tool for the job. Sometimes your goal is to get the widest possible dissemination of your code, in which case you might choose a very permissive license such as BSD. Other times you might want to encourage a lot of downstream sharing and coherence, so you might find the GPL more appropriate.

"Free" Software Licenses

The development and use of free software is overseen by the Free Software Foundation (FSF), established in 1985 to promote computer users' rights to use, study, copy, modify, and redistribute modified computer programs. The FSF supports many free software projects, but particularly the GNU Project and its GNU operating system. The GNU Project was conceived in 1983 as a way to bring back the cooperative spirit that prevailed in the computing community in earlier days—to make cooperation possible once again by removing the obstacles to cooperation imposed by the owners of proprietary software. The GNU GPL is a backbone of the free software movement.

Note that the word *free* in "free software" pertains to freedom of access and freedom from onerous restrictions, not price. You may or may not pay to get free software. Either way, once you have the software you have three specific freedoms in using it: (1) the freedom to copy the program and give it away to your friends and coworkers; (2) the freedom to change the program as you wish, by having full access to source code; and (3) the freedom to distribute an improved version and thus help to build the community. (If you redistribute GNU software, you may charge a fee for the physical act of transferring a copy, or you may give away copies.)

The following licenses qualify as free software licenses, and are compatible with the GNU GPL (for detailed information and comments about these and other free software options refer to www.gnu.org or http://gplv3.fsf.org/):[2]

- GNU GPL Version 3

- GNU GPL Version 2

- Apache License, Version 2.0

- Artistic License 2.0

- Berkeley Database License (a.k.a. the Sleepycat Software Product License)

- FreeBSD License

- Intel Open Source License

- License of Netscape JavaScript

- OpenLDAP License, Version 2.7

- License of Perl 5, and earlier

- License of Python 2.0.1, 2.1.1, and later versions

- License of Ruby

- X11 License

- License of zlib

The following licenses are free software licenses, but are *not* compatible with the GNU GPL (a module covered by the GPL and a module covered by the licenses in the following list cannot legally be linked together; in many cases, these licenses have a newer version that may be GPL–compatible):

- Academic Free License, all versions through 3.0
- Affero General Public License Version 1
- Apache License, Version 1.1
- Apache License, Version 1.0
- Apple Public Source License (APSL), Version 2
- Original BSD license
- Common Development and Distribution License
- Common Public License Version 1.0
- Condor Public License
- Eclipse Public License Version 1.0
- IBM Public License, Version 1.0
- Interbase Public License, Version 1.0
- Mozilla Public License (MPL)
- Netizen Open Source License (NOSL), Version 1.0
- Netscape Public License (NPL), versions 1.0 and 1.1
- Nokia Open Source License
- Open Software License, all versions through 3.0
- OpenSSL License
- Phorum License, Version 2.0
- PHP License, Version 3.01
- License of Python 1.6b1 through 2.0 and 2.1
- Q Public License (QPL), Version 1.0
- Sun Industry Standards Source License 1.0
- Sun Public License
- License of xinetd
- Zope Public License Version 1

Nonfree but Free-Sounding Software Licenses

The following licenses *do not qualify* as free software licenses as defined by the FSF. A nonfree license is automatically incompatible with the GNU GPL. Of course, we urge you to avoid using nonfree software licenses, and to avoid nonfree software in general. There is no way we could list all the known nonfree software licenses here; after all, every proprietary software company has its own. We focus here on licenses that are often mistaken for free software licenses but are, in fact, *not* free software licenses:

- Apple Public Source License (APSL), Version 1.x
- Artistic License 1.0
- AT&T Public License
- eCos Public License, Version 1.1
- GPL for Computer Programs of the Public Administration
- Hacktivismo Enhanced-Source Software License Agreement (HESSLA)
- Microsoft's Shared Source CLI, C#, and Jscript License
- NASA Open Source Agreement
- Open Public License
- Reciprocal Public License
- SGI Free Software License B, Version 1.1
- Squeak License
- YaST License

A Closer Look at the GPL

At first glance the number, variety, and requirements of the various licenses appear somewhat daunting. For 95% of you, however, there's a short list of licensing options with which you should become better acquainted: GPLv2 and GPLv3, BSD, Apache, Mozilla and derivatives, CDDL, and Creative Commons. So, allow us to add some contour to the descriptions and comments provided previously, focusing, for now, on the GPL.

Think of the GPL as simply a license that is appended to the source code itself. It says, "Look, you can do anything you want with this code. You're free to make copies of it, to make improvements to it, and to use it commercially in the way you see fit, but you have to play by the rules." And there are some really important rules. One is that you've got to propagate the license. You can't just strip out the code and put it into your own product. Another is that if you fix bugs or make improvements, you have to make those changes freely available. You have to publish that source code and put it back into the community under the same license.

The license can also cover patents that might be on the code. (GPLv3 makes this explicit—there is ambiguity in GPLv2; CDDL is also explicitly granting.) If you play by the license rules, the person who wrote the code gives you a patent grant, saying, in essence, "If I have any patents that would allow me to exclude you from expressing that idea, meaning that they're relevant to that particular source code, I grant you those patent rights." In short, I won't prosecute you as long as you follow the other rules in the license. Now, if you decide not to follow those rules, the patent grant is rescinded, and I can come at you on those grounds. Furthermore, if you use the code you offer a reciprocal grant for any patents you might hold. If you then decide to prosecute users of the code on the basis of your patents, you are also no longer playing by the rules and you lose all rights to the original code. It's a kind of "mutually assured destruction" that is intended to significantly dilute the value of software patents.

The GPL has another aspect to it, as we alluded to earlier—a viral aspect— that is the subject of much debate in the open source community. Basically, it says that if you have a piece of code that is in the GPL, code that it touches (and much debate centers on what is meant by *touches,* but technically it is part of an "aggregate") must also be licensed under the GPL or a permissive license that is compatible with the GPL (e.g., BSD or Apache). People use a lot of techniques to work around this restriction, such as library interfaces and dynamically loadable modules. But the spirit and intent of the license are clear: If you use GPL code, you should be working toward, not against, software freedom.

To be clear: You can't take code under the GPL that you haven't written and place it under another license. And you don't get a patent grant for any of the ideas expressed in that code that you choose to recode under a different license.

It's important that you examine the key differences between GPLv2 and GPLv3, and for that we recommend "A Quick Guide to GPLv3," by Brett Smith, **available at** www.gnu.org/licenses/quick-guide-gplv3.html. Version 3 includes a number of improvements to make the license easier for everyone to use and understand. It includes stronger protection against patent threats; it offers more ways to provide source code to users; and it is a more globally applicable license. But even with all of these changes, GPLv3 isn't a radical new license; it's an evolution of the previous version. Though a lot of text has changed, much of it simply clarifies what GPLv2 said.

So, whether you choose GPLv2 or GPLv3, by putting something under the GPL you have a lot to say about what can and can't be done with it. Why, then, are there so many other open source licenses, such as Apache, Mozilla, and CDDL? Because the GPL is sometimes considered too restrictive in its terms. A key difference with the Mozilla license, for example, is that it removes the viral requirement and permits code of different licenses to be commingled (assuming that other licenses permit the commingling as well).

This commingling turns out to be really key for situations when you don't control all of the modules in your system or when you want to work off an open base but do some of your own extensions. And yes, that does go against some of the "software freedom everywhere" philosophy implied by the GPL, but it helps to strike a balance between a whole set of competing interests.

Open software is fundamentally about developer freedom. Complementary to developer freedom are developer rights. A code developer (an individual or a corporation) does have rights to the code he developed. It is, after all, the fruit of his labor. By choosing to place that code under an open source license, a developer surrenders some, but not all, of those rights to the community in the hopes of a beneficial exchange: No open source license gives away all rights.

Contributor Agreements

When starting an open source project, the choice of license is intended to be permanent. One question not really considered by open source licenses is how to handle *multiple licenses or relicenses* of the code. There are many reasons why a new or alternative license might be considered. The Linux kernel, for example, is covered under GPLv2, but suppose Linus Torvalds wanted to move it to GPLv3.[3] Linus couldn't do that unilaterally because he is not the sole copyright holder on the kernel code modules. Just like you can't rip someone else's code and put it under a license you choose, you can't relicense it either, unless you get *all of the copyright holders to agree*.

Without some sort of aggregated copyright, every single contributor must be contacted and unanimity reached in order to relicense a code base, or parts of the code must be reimplemented. This is true for all but the most permissively licensed open source projects.

There are other reasons why multiple licenses might want to be maintained. Suppose that a person or company (say, some sort of electronic equipment manufacturer) wants to use your open source project in its device, but it finds the rules of the license too onerous. For example, it might want to make some improvements or additions as points of differentiation, and keep them secret. One option is that the company could *pay you* to give it a version of your code under a proprietary commercial license. That is, you would get compensated for removing the open source license obligations from the code. This Original Equipment Manufacturer (OEM) model is a fairly common one for companies that are trying to build a business around open source software.

A fundamental governance issue for an open community is how the copyright (and other interests) for contributions is maintained.

Contributor agreements (CAs) are used by many companies, open source communities, and other organizations to set forth the terms under which contributions can be made to a project. Sun's CA, for example, is the contractual vehicle for contributions to Sun open source projects such as OpenSolaris, OpenJDK, and GlassFish.

A CA can be a source of clarity about the terms of a relationship—or a source of confusion and heartburn. Here are our answers to frequently asked questions; we hope they'll help you understand the key considerations of signing a CA and how a CA benefits both software engineers and commercial enterprises.

What Does the CA Do?

Typically, when you sign a CA, you agree to share your copyrights with a specific governing body (a company, community, or other organization) and license any patents bearing on your contributions to that organization. You are asserting that your contributions are original works; that you are legally entitled to grant the organization these rights; and that your contributions do not violate anyone else's rights. By accepting a CA, the organization promises that your contributions will remain free and open source software (i.e., the software will be published and will remain available under a free or open source software license). But typically, that is not the *only* license under which the code could be made available. Specifically, a CA does the following.

- It allows an organization to sponsor projects while retaining the ability to offer commercial licenses. Without this ability, a commercial enterprise could not responsibly open-source code bases that in some cases represent hundreds of millions of dollars of shareholder investment.

- It allows the company to protect community members (both developers and users) from hostile intellectual property litigation should the need arise.

- It protects the integrity of a base of code, and in turn it protects the community around that code base: the company, the developer community, and the project's users. In Sun-sponsored projects, for example, Sun acts on the community's behalf as a steward of the code in the event of any legal challenge. This is in keeping with how other code stewards, such as the FSF, defend projects. In order to represent a code base against legal challenges, Sun must have copyright ownership of all the code in that project. Consolidated copyright of code also allows for the possibility of relicensing the whole code base should that become desirable.

- It allows for simpler relicensing. When starting an open source project, the choice of license is intended to be permanent, but the experience of the past few years is that the ability to relicense a project is a useful tool in meeting challenges to free and open source software (and especially challenges from the proprietary software market), and not having that flexibility may be a drawback. Without an aggregated copyright, every single contributor must be contacted and unanimity reached in order to relicense a code base, or parts of the code must be reimplemented. This is true for all but the most permissively licensed open source projects.

- It allows the sponsoring organization to act as a bridge between different communities using the same code under different licenses. This, in turn, allows the sharing of code among open source projects that might otherwise not be possible, and it allows the company to license source code to parties who are not yet prepared to work with an open source license.

The joint copyright assignment also gives the original donor of the code base the ability to offer commercial, binary distributions of the project. Without this ability, it would not be possible for commercial enterprises like Sun to open their technologies.

Do I Lose the Rights to My Contribution by Signing a CA?

No, a CA should ask you only to share your rights. Be wary of a CA that requires you to transfer copyrights to another organization. When you agree to a Sun Contributor Agreement (SCA), for example, you grant Sun joint ownership in copyright, and a patent license for your contributions. You retain all rights, title, and interest in your contributions and may use them for any purpose you wish. Other than revoking the rights granted to Sun, you still have the freedom to do whatever you want with your code. Sun may exercise all rights that a copyright holder has, as well as the rights you grant in the CA to use any patents you have in your contributions. As the SCA provides for joint copyright ownership, you may exercise the same rights as Sun in your contributions.

Note that a CA does not necessarily give you a say in the relicensing of your contribution and the use of your granted patent rights. Nor can you be certain that your contributions will make their way into the products and distributions that the company actively markets, or that they will be used only for the advancement of free software. However, through the governance processes for each project and community, participants usually have a strong voice in how the code base as a whole evolves. Consult the governance policies of the projects to which you contribute for specific details.

Will I Receive Credit for My Contributions?

This is up to the company that originates the CA. A CA does not obligate the organization to offer any particular form of credit or recognition for contributions; such policies are determined by individual projects. You should consult a specific project's governance and license documentation for more information.

Can I Contribute the Same Works to Other Projects?

Assuming the CA for a project allowed for shared rights, if you contribute to that project you will retain the right to contribute to other projects not sponsored by that organization under any license. Remember, you are asked only to share rights, not to relinquish them. Contribution policies of other projects to which you might want to contribute may restrict your ability to contribute works you've contributed to an organization's project, or to participate in some roles if you have participated in a project at that organization. Consult their policies for more information.

When Should I Sign a CA?

You'll be asked to execute a CA *before* you make any contribution—no matter how large or small your contribution might be. The requirement for a signed document is an unfortunate consequence of copyright law in some jurisdictions. Here are a couple of guidelines.

- If you'll be contributing on behalf of a company, an officer of your company (usually a VP or higher title) must sign the CA on behalf of the company.

- If you've previously assigned copyright in your prospective contribution to an open source organization (e.g., the FSF) under its contribution policy, you no longer have the ability to assign a joint copyright to a company. However, the open source organization will probably have granted you back an unlimited, sublicensable copyright license to your contribution, and other accepting organizations may also grant back such a license. This kind of grant-back copyright license may allow you in turn to grant to a company all the rights needed under the CA.

You can stop your participation in a project at any time, but you cannot rescind your assignments or grants with respect to prior contributions. This protects the whole community, allowing downstream users of the code base to rely on it. The company originating the CA typically cannot terminate its responsibilities under the CA either.

What if I'm Working for a Company but Contributing as an Individual?

The lines between "employee" and "individual contributor" continue to blur. Let your employer know whether you intend to contribute to an open source project—whether as an employee or as a private individual—and consult the company's IP specialists if there are any potential issues (see also the discussion on employment contracts in Chapter 12).

Software Indemnity

As the popularity of open source software grows, so do concerns that companies and even individual engineers might be risking a lawsuit if they build or run applications that infringe on someone else's IP.

The ghost of the 2003 patent-infringement lawsuit brought by SCO against IBM continues to scare people. SCO alleged that IBM moved technology from UNIX to Linux against the terms of its contract with SCO, violating patents and trade secrets in the process. SCO sought $5 billion from Big Blue, and also tried to compel Linux-using corporations to license SCO's UNIX.

SCO lost the battle and the war (the company is no longer a going concern and a judge has declared that there is no UNIX in Linux), but the fear remains. Since open source licenses declaim liability for SCO-style attacks, and since most open source software projects aren't exactly flush with cash to pay settlements or lawsuit judgments, open source software is often perceived as riskier for companies than proprietary software would be.

Yes, you can get indemnification for most open source software packages. There are now companies that specialize in this type of "insurance," such as OpenLogic, which provides indemnification coverage for an extensive list of open source packages (as long as you've purchased open source technical support from OpenLogic). Other companies have taken steps to indemnify developers: Red Hat and Canonical both offer their fee-paying customers indemnification for SCO-style threats, and Sun, as we mentioned in Chapter 12, has paid dearly to indemnify the entire Java community against infringement claims made by Kodak.

But the question remains: Is software indemnification really necessary? And does your company need to pursue an indemnification strategy for the software innovations it creates? According to Stephen O'Grady, industry analyst for RedMonk,[4] the risk of prosecution for illegal use of open source software is similar to that of being sued for using proprietary software—and they're both "exceedingly low," from an historical perspective. His opinion is that open source is not uniquely vulnerable to patent claims, and that "no software, open or closed—is risk-free." He adds that he feels it is not a feature for which it is worth paying a premium or altering a buying decision.

14

Creativity and Control

"Practically speaking, copyright and patent and trademark law have only one thing in common: Each is legitimate only as far as it serves the public interest. Your interest in your freedom is a part of the public interest that must be served."

—Richard Stallman,[1] founder of the GNU Project

In the previous two chapters, we took a pragmatic approach to intellectual property (IP): Understand how the existing system of protections works, and then leverage that system to create a range of techno-business models, from narrowly controlled (proprietary) to globally participative (open). We hoped to show the range of possibilities without being overtly biased about recommending a particular path. And we avoided posing basic questions such as whether the existing system of protections is best for fostering innovation.

In this chapter, we will definitely let our biases percolate through. If we are successful, you should leave this chapter understanding our fundamental belief that in this new era, innovation is best served by structures that promote openness and sharing—that there is a dynamic tension between creativity and control, and that by easing back on the controls, creativity will become even richer.

This viewpoint is, of course, not original to us. We've been very much influenced by the writings of Lawrence Lessig,[2] who in turn has stood on the shoulders of pioneers such as Richard Stallman, Eric Raymond, Mitch Kapor, Eben Moglen, and Linus Torvalds.

Maximizing the Cycle of Innovation

The instruments of control—from patents, copyrights, and trademarks, to digital rights management (DRM) technologies—seem to be at odds with creativity. After all, these controls are all about *excluding* or *restricting* others from using an idea, or copying works and content. Yet very few creative works, including software programs, are truly original. We invariably build upon the works of others. As filmmaker Martin Scorsese said: "The greater truth is that everything—every painting, every movie, every play, every song—comes out of something that precedes it.... It's endlessly old and endlessly new at the same time."

Put another way, most creative works are in fact derivative ones. They incorporate ideas (and their expressions) from others and then build, reinterpret, improve, optimize, and ultimately return better ideas and better-engineered artifacts. If the instruments of control are set right, this virtuous cycle continues, and a rising tide of innovation lifts all boats.

Moore's law is a superb example of the innovation cycle—so much so that we can accurately predict that a key innovation metric of semiconductors (transistors per chip) will indeed double about every eighteen months. This is not a law of physics; it is a law of techno-economics. Fifteen years ago, for example, very few people could predict exactly how we would stay on this exponential cycle, and yet ten doubling times later we have about a thousand times (!) more transistors per chip.

Evidently, our system of intellectual property controls in the business of semiconductor manufacturing is having some positive effects. Companies all along the value chain, from equipment suppliers to fabricators, continue the enormous cycles of R&D and capital investment that manifest Moore's law. And there are similarly breathtaking capital cycles in the pharmaceuticals industry, with $1 billion spent on bringing each new "molecular entity" (drug) to market, meaning tens of billions of dollars are spent in R&D each year.[3] Intellectual property controls play an essential role in this industry to exclude others from the much easier task of replicating a compound that took many years of forward investment to discover and test.

A trap to avoid is generalizing from these long-term, very capital-intensive innovation cycles, and concluding that everything is working as it should. The question we should always be asking ourselves is whether we have struck the right balance:

"Does our system of controls maximize innovation fairly?"

The *fairly* qualifier deserves some explanation. By this we mean fairness in both access to markets by competitors along with fairness in compensation for one's creative efforts. Overcontrol can damage both senses.

Our view is that our system of controls is far from the best one. Like Lessig, we believe a better balance can and should be struck along the control-creativity continuum that serves the interests of all: engineers, businesses, and the public. We view Free and Open Source Software (FOSS) as the harbinger of systems that unleash very rapidly moving innovation cycles through sharing and global participation.

To be sure, extraordinary FOSS innovation has happened under the current system of intellectual property controls. The genius of Stallman and Moglen with the Free Software Foundation (FSF) and Software Freedom Law Center, and Lessig with the Creative Commons, has been to parlay the existing copyright laws into a whole new set of concepts regarding the licensing of IP. As described in Chapter 13, this is accomplished by writing a new license (e.g., an "open source" one) that is incorporated into copyrighted material. If you don't follow the terms of the license, you run afoul of the existing copyright law.

Perhaps our existing system is malleable enough to get us to a better balance. Perhaps. We think there are many opportunities for reform, from the issues of software patents and patent trolling to the absolutism of control over interfaces and DRM.

But no matter your view on the necessity of reform, there are plenty of ways you can be progressive and tap into the tremendous power of creative, freely moving global innovation networks.

How We Got Here

In Chapter 12, we provided a primer on the primary landscape of intellectual property law as it exists today: patents, copyrights, trademarks, and trade secrets. Here, we'll provide a bit of context as to why we ended up with our current system as a segue to the discussion of what could be next.

First, let's back up and look at the term *intellectual property*. It's a loaded one. Many people—the authors of this book included—bristle at it because it evokes the idea of real property, which it clearly isn't. We also surround the term with analogues of real property: People "steal" software or music; they become "pirates." Promoters of participation and sharing get labeled as "communists" (we assure you, the book's authors are most definitely capitalists).

Why is IP unlike real property? Because the consequences of its misappropriation can be wildly different. Think of an apple (the fruit). The apple is real property. If I give or sell my apple to you or you steal it from me, you have one and I'm deprived of one in a very direct and measurable way. Same thing if you steal my vinyl record of *Led Zeppelin IV*. However, if you "steal" a copy of my digital music, I am not deprived of anything other than the money you might have paid, and *I still have my music*.

And the damages for IP infringement are often incongruent. If you dump toxic sludge on my front lawn, you have to pay something proportional to what the real property damage was (oh, and maybe something punitive on top). When you get punished for patent infringement, the settlements can be surreal. A patent holder, for example, can extract hundreds of millions of dollars for a patent he may have paid thousands for, and whose infringement causes him no direct harm whatsoever.

With our discomfort with the term *intellectual property* duly noted, let's not get bogged down over it. Protection mechanisms for IP seem to have always had a role in human cultures. Look at any era and you'll find examples. In ancient Greece, inventors of new recipes were granted an exclusive right to make that food for one year. The Republic of Venice started issuing patents in 1474, protecting inventors by prohibiting others from replicating their new devices.[4] The United States has had a copyright law since 1787, and in 1884 the U.S. Supreme Court concluded that photographs could be copyrighted.

Moreover, intellectual property law has long played a role in shaping the development of key innovations—sometimes encouraging new ideas, sometimes inhibiting new ideas, sometimes encouraging new ideas by getting in the way. Here are some oft-cited examples from the history of patents in America.

- In 1781, James Watt, who built the first full-scale steam engine, was forced to circumvent a patent held by James Pickard in order to make his engine's flywheel turn—and as a result his team invented the "sun and planet gear" which played an important part in the development of devices in the Industrial Revolution.[5]

- Between 1878 and 1892, the electric light industry was growing in terms of installed lights but shrinking in terms of competition as both Thomas Edison and George Westinghouse determined to control the industry and its advancement. They formed the Board of Patent Control, a joint arrangement between General Electric and the Westinghouse Company, to defend the patents of the two companies in litigation. This proved to be a wise decision as more than 600 lawsuits for patent infringement were filed.

- The development of radio technology was blocked by the patents held by various individuals and corporations, and it wasn't until the U.S. government stepped in and forced cooperation among various competing patent-holders that the technology moved forward.[6]

Students of the field (and the professionals who practice it) would all point out that these are examples where the patent system "maximizes innovation" by either sparking new ideas in an effort to avoid existing patents or encouraging the pooling of IP among competitors, thus advancing collective innovation.

Let's look at this latter aspect in more detail. Big patent exchanges or cross licenses among companies are often seen as signs of a healthy system. One of the most visible consortia is the MPEG LA[7] which is a clearinghouse licensing authority for a plethora of patents relating to the MPEG video and audio encoding formats (MPEG-2 and MPEG-4, in particular). If you wanted to, say, build a new set-top box that would plug into the Internet and play MPEG video streams on a TV, you'd be advised to obtain a license from MPEG LA. Expect to pay several dollars per box for the pleasure[8] which might be about the same amount of money in net profit you were planning on.

But here it is, all in one place. A set of intellectual property that enables the worldwide encoding and playback of digital media. That would seem to be an ideal outcome: an opportunity for great technology to be developed and standardized and a way for compensation to flow back to the inventors. Certainly such a system is best for innovation, isn't it? In our view it isn't. A better system is one that would enable a much freer flow of ideas and implementations.

From another vantage point, licensing the intellectual property around broadly adopted interoperability standards is mostly just a tax. But more importantly, the tax rate can be set arbitrarily, even much higher than the intrinsic value of the underlying innovation. The thing that makes a standard valuable is that *it is the standard*. In video encoding, for example, there are lots of alternatives, many of which are technically superior to the MPEG suite. So, the value of the MPEG suite is that we all agree upon it. And by agreeing upon it, we get to create a very strong network effect. That network effect—the consumer benefit of being able to view any video on any player—becomes incredibly more valuable than, say, a new algorithm that compresses at lower bandwidth with higher quality.

To have IP claims come in *ex post* (after the standard has been agreed to) is unfair at best and mostly just plain exploitive. At a minimum, we strongly believe that IP claims surrounding a potential standard should all be aired *ex ante* (before the standard is agreed to) and Reasonable and Non-Discriminatory (RAND) pricing exposed so that at least there can be an informed discussion about whether a particular bit of innovation is actually worth the license fees. We have little doubt that if we were to move to this level of transparency, many other less-expensive alternatives would surface.

Ultimately, we are convinced that royalty-free alternatives would emerge, and these would be strongly preferred. Why? Because in this case, the system becomes open to cycles of improvement that can come from anywhere. There is no monetary franchise that is being protected by keeping the world attached to a stagnant technology that generates income for the promoters of the standards in the first place. Furthermore, the downstream innovations also open up because many more people can figure out ways to use the standardized technology in new and unanticipated ways.

None of this discussion around IP as related to standards should be construed as an argument that we should abandon the patent system. We are pointing out a particularly (in our minds, at least) anti-innovation consequence of when the intrinsic value of an invention is nowhere close to the network value that can be created, especially through the process of creating standards. It's an area in which we all should proceed very carefully.

"We live in a world now that consists of pipes and switches," said Dave Crossland in his legendary talk at ITC-ILO in Turin back in December 2004. "Pipes that move things from place to place, frictionlessly, at the speed of light. And switches that determine who gets which things, when, how, with what control, and at what price . . . and the rules that they use to determine who gets what, when, where, how, and at what price, are computer programs. Those who control computer programs, control who gets everything."

Control over Interfaces

The issues that arise concerning IP and standards are actually a special case of a bigger concept: controlling markets by controlling interfaces. By *interface* we mean an essential agreement (mechanical, electrical, logical) that connects things. A simple example is a razor handle and razor blade cartridge. The interface is the mechanical linkage between the two: the way the razor cartridge clicks into the handle.

The point we make here is that the interface is an essential feature to control. If, for example, you obtain a patent on that mechanism, you can effectively *exclude anyone else from making a compatible blade.* That is extraordinarily powerful. You can completely control the marketplace for one thing (the blade) by controlling the way it connects to another (the handle). And this supports a key business model; give away the handle and charge (handsomely, as it were) for the blades.

This same model applies to ink jet printers and cartridges. Ink jet printers are almost literally given away because the proprietary ink cartridges are very profitable. Protection of intellectual property (especially patents and trademarks) can apply to the printer and to the cartridges, including printing mechanisms and ink formulations. But the jewels are the patents that cover the way the cartridges mechanically and electronically mate to the base printer assembly (the interface).

You can also apply a patent to the way electrical contacts are made. Figure 14–1 depicts a photo of an electric motor on the left and a plug on the right.

FIGURE 14–1   Deciding What to Patent

the right. Assuming both were new, on which device would you choose to get a patent? If you owned the plug, you could exclude (or tax) all possible innovations in motors. Or you could choose to keep the motor market for yourself.

There are many other examples of interface control. A microprocessor manufacturer that obtains a patent on some details of a chip socket or pin signaling can effectively exclude other chips from plugging into motherboards that support that socket. And this is true even if there is no infringement by the chip itself and the motherboard is freely available from others in the market. The encoding of data sent over the Internet (say, a video encoding scheme) and a protocol for accessing a service are other examples.

Clearly interface control is powerful. The limited monopoly afforded by a patent is amplified when an interface is involved. Essentially, the right-to-exclude concept, when applied to an interface, has the effect of *excluding all possible implementations*.

No doubt you can draw a few conclusions from this discussion. A purely business-motivated one could drive you to want to identify and control interfaces in your own engineering domain. After all, monopolistic profits are by far the richest ones. Another conclusion, and one we hope might just be tugging at your Citizen Engineer side, is that perhaps this is anti-innovative. Yes, if *you* happen to be the lucky (or smart) one to enjoy such powerful exclusive rights, you are likely to be pretty happy with the outcome. But if you want to enter the market with a new idea and are categorically excluded from doing so, you will think differently.

Most importantly, if you are a consumer or customer of the technology, you likely are paying too much and having your choices limited. That is, the societal interest in a system of controls that maximizes innovation fairly is not being well served. If you think of yourself as a Citizen Engineer this should bother you. We are using our skills to design things, and our societally afforded system of IP controls acts in a way that might not be in the best interest of society.

Happily, there is something you can do. There is a basic set of concepts and principles that you can adopt that can align your interests with those of society. There are choices you can make in the way you approach the balance of creativity and control. We'll tell you how.

Innovation Commons

In his extraordinary book *The Future of Ideas*,[9] Lawrence Lessig identifies and explores the concept of an *innovation commons*, using the Internet as the exemplar of how intense innovation can flourish globally when the right balance between freedom and controls is struck. The book is also presciently cautionary about how accumulation of control points, often facilitated by our IP system, can throw a serious amount of sand in the gears—or worse.

Lessig describes how various layers can be composed, some that are proprietary (the physical infrastructure of the Internet is almost all privately owned) and other key layers that are free (e.g., the TCP/IP protocols for the communications layer and the HTTP and HTML standards for the Web layer). It's the mixture of these, and particularly the ability of the free layers to prevent monopolization of the private layers, that creates this globally accessible place for what is, by any measure, an enormous cycle of innovation.

There are also growing infrastructure and application layers in the Internet that are creating an even richer commons. The Linux operating system has been fundamental to the growth of the second generation of the Web. Also key have been free and open source application frameworks and services, notably the Apache suite of Web servers and the MySQL database.

The word *commons* is meant to evoke something akin to its figurative use: "The mutual good of all; the abstract concept of resources shared by more than one, for example air, water, information."[10] An innovation commons is an engineered place where we decide to collectively and freely contribute ideas and their expressions (e.g., specifications or code). By freely contributing, and then often collectively evolving, enabling interfaces and implementations, we get to do a number of things. Most important is that we serve society's interest by providing a fair place for innovation to thrive. That in turn serves our own business interests by providing even more opportunities to build and be compensated for implementations that are superior and/or add value on some important customer dimension.

Our use of the word *free* deserves attention. The freely available layers in an innovation commons are—to use Stallman's words again—"free as in freedom."

They might, and oftentimes are, "free as in free beer," but that is simply a technique to maximize adoption and participation by lowering barriers (financial ones). The freedom aspect is the key one. It's about the ability to *freely build upon each other's works,* the ability to share and innovate on top of, alongside, or in areas not originally anticipated. Again, and a central point here, is that the "free" layers aren't given away; they are contributed to the commons and are free in that context.

Open licenses play a defining role.[11] But they are just the beginning of the formation of a community. Eben Moglen captured the essence of this in his 2007 OSCON address, on the "Republic of Open Source." Here are Moglen's thoughts as transcribed by blogger John Eckman:[12]

> *Licenses are a part of community building—they are constitutions for communities. But the words of licenses are just the beginning—just as written constitutions are just the beginning of the republics which they give birth to and organize. In the 21st century it is no longer factories or individuals which are the units of production and distribution—it is communities. This is the reality of mass culture.*

We'll go into more detail in a moment on the secrets of keeping a community healthy. But properly ignited and maintained, the communities around an innovation commons have the potential to greatly expand the available market for all; a place of greater efficiency, faster cycle times, and competition. If we work together as Citizen Engineers to discover and build such free areas, we simultaneously serve market opportunities and society. We think that's a good thing.

The Economics of Open Source

We said earlier that mastery of intellectual property law is a great way to control the destiny of ideas—it's a powerful lever for getting people to participate in your ideas, and for using the ideas of others responsibly. But how does all that translate to better business results?

Richard Stallman was one of the first people to recognize that one great way to get people to participate with your stuff is to make your stuff free. "That's free as in freedom, not free beer," as he puts it. Basically, what Stallman was trying to do was to create open development communities: Others can take what you've done and they're free to build upon it and create new things, so you become part of something bigger. And so do they.

Great. But how do you stimulate that, and how do you make money at the same time? Let's look at the example of open source software. The thesis is that the open source model creates communities; communities create markets; then you can go in and monetize markets.

Red Hat is a great real-world example. This is a company built on a free software platform—a platform it didn't create. The software is available free of charge. Service and support will cost you—and Red Hat has built its business on support.

Or consider MySQL, the world's most popular open source database, used by virtually every Web 2.0 and e-commerce company today. The company had an extremely attractive *business* model (when we bought it in January 2008). MySQL had built a vibrant developer community that was millions strong and was known for a passion for disruptive innovation. Tapping into that community allowed Sun to enter new markets, drive new adoption of its hardware products, and open doors and create new opportunities for ISV partners and channel partners. It's a win for innovation and a win for business. They're not mutually exclusive.

Why is there a booming market for support of "free" software? Because at the enterprise level, every CTO wants to use free software, and no CIO will allow the use of free software without a commercial support contract. Not one. Downtime is just too expensive—millions of dollars per minute in some cases. So, they're smart to make sure they have someone standing behind the products they use to run their businesses.

On the other hand, if you can try a product before investing any money in it, why wouldn't you? The beauty of open source software is that customers can install it on their test servers free of charge, run it through any tests they like, and decide for themselves whether the program is all it's cracked up to be. They don't have to take a salesperson's word for it. Then, if they decide to deploy it, they can come back to you for a support contract.

There's no reason you can't offer a traditional commercial license as well. In fact, for competitive reasons, very few Internet services or embedded products companies will agree to reveal their code—a requirement with some of the open source licenses we discussed earlier. Such companies will gladly pay for a commercial license—and a support contract.

We believe open source is one of the most significant developments in network software collaboration. The world of network computing is bigger than any single company. It's a world that really requires the coordination and interaction and building of ecosystems that go across our entire industry and involve many, many people in many, many worlds. If you think about what we need in order to come together to coordinate, to make things happen, it's this community.

Beyond Software

Most of our reasoning about creativity and control, and the power of collaborative communities, is shaped by what has happened in software and, more broadly, the Internet. We certainly take liberties in extrapolating from open source software and inferring a more fundamental state for engineering and innovation: a freer state where innovation happens everywhere.

A fair question is whether such extrapolation is well founded. Could it be that there is *something special or different* about software development that lets that branch of engineering thrive under collaborative development, but not other branches? You might guess that we think what has happened in open source software is a leading indicator, a template as it were, for most every other branch in the collective art of engineering.

Why it started with software is easy to explain, though the reasoning is a bit circular. The Internet is the key *enabler* of global collaborative communities because it drives the cost of communication and sharing to near zero and because it can readily scale so that the essential network effects in these communities are allowed to take root and compound. We shouldn't be too surprised that the discipline at the foundation of the network itself, software engineering, is the first to take advantage of it. There also just might be a bit of good luck and fortune that the pioneers of the Internet came from an academic and research culture of extramural collaboration.

But again, perhaps there is something special about software and it simply doesn't generalize to other disciplines. We'll explore that concern here through a series of examples.

Goldcorp

In their book *Wikinomics: How Mass Collaboration Changes Everything,*[13] Don Tapscott and Anthony Williams chronicle the remarkable story of a struggling Canadian gold producer, Goldcorp, and its failed internal attempts to convert a wealth of geophysical data that suggested that it *should* be able to get more gold into actual production. The data suggested perhaps 30 times (!) the amount it was actually mining. But the precise interpretation of the data eluded corporate geologists throughout the 1990s.

In frustration, Goldcorp's CEO, Rob McEwan, decided on a radical path. He had just attended (in 1999) a conference at MIT on open source software, and wondered whether the same type of community participation model would work in the geophysics industry. His idea was to take all of Goldcorp's proprietary geophysical data accumulated since 1948—about half a gigabyte covering

55,000 acres of the company's Red Lake property—and make it freely available on the Internet. Moreover, he decided to fund what was called the "Goldcorp Challenge" with $575,000 in prize money, awarded to those who contribute to the discovery of new sources of gold on the property.

The community response was stunning. More than a thousand participants, ranging from practicing geologists to students to military officers, from more than 50 countries tried their hand at the data. A whole slew of new approaches and techniques were brought to bear, well beyond what had been attempted by Goldcorp's staff. In the end, 110 targets were identified, more than 50% of which were new. About 80% of the new targets yielded what ended up being 8 million ounces of gold. Goldcorp has subsequently become one of the most valuable companies in the industry today.

This style of bounty for community participation is being tried in a number of different settings. Notable is Google's "Android Challenge," which provided $10 million in awards for developers who built great applications for Android, the first open platform for mobile devices.*

TCHO Chocolates

TCHO is a San Francisco-based start-up that is attempting to revolutionize the way "obsessively good" dark chocolate is made using a range of open community strategies. At the consumer end, aficionados are invited to be "beta" testers of new recipes that are being developed in TCHO's laboratories. One of the key goals is to establish a flavor system, the TCHO Flavor Wheel, that lets consumers much more accurately reflect their taste preferences. The testing and evaluation are done via the Web in a highly participative manner. Both consumers and TCHO benefit from this; chocolate lovers have greater confidence that they will be buying something they will really enjoy, and of course, TCHO gets a much deeper understanding of what people will want to purchase.

Involving your customer via the Web is certainly not remarkable, but it is the next step in community building that makes TCHO a most interesting experiment in an innovation commons. TCHO is opening up the Flavor Wheel for participation by the cacao bean farmers themselves, many of whom have apparently never tasted chocolate nor know what the manufacturing process is like; as such, the farmers have been simply growing beans to get the highest yield against a fairly rudimentary system of quality grading. These beans

---

* Android Challenge is a contest sponsored by Google in which $10 million in prizes was awarded for mobile applications built on the Android platform (see http://code.google.com/android/adc.html).

get packed up and sold to wholesale chocolate producers, who in turn resell the bulk chocolate to "re-melters" who form most of the common brands. The current state of quality grading is likened to labeling a wine "France, 13% alcohol."[14]

Under the Flavor Wheel system, cacao farmers (for whom TCHO has provided Internet connectivity, incidentally) can see the market value of different kinds and qualities of cacao beans, and they can bias their production to where they can maximize their profits. TCHO is hoping the farmers form an open community of their own, where increasingly better methods are being tried and exchanged.

The Open Prosthetics Project

The concept of free and "open source" can apply to all forms of engineering, and mechanical design is no exception. Combine this with a core aspect of networked communities—the ability to bring together people under a common cause completely independent of whom and where you are—and you get some truly remarkable outcomes. One of these is the Open Prosthetics Project (OPP), which is part of an organization called the Shared Design Alliance. Their Web site explains:[15]

> The Open Prosthetics Project is producing useful innovations in the field of prosthetics and freely sharing the designs. This project is an open source collaboration between users, designers and funders with the goal of making our creations available for anyone to use and build upon. Our hope is to use this and our complementary sites to create a core group of lead users and to speed up and amplify the impact of their innovations in the industry.

A central focus of the OPP is the reengineering of a widely used, but no longer manufactured, arm prosthesis called the Trautman hook (which was introduced in 1925). Participants in the project contribute in all manner of ways (from designs, to testing, to ideas) in an effort to improve upon the original design. The "open source" aspect of this project is that these ideas can be freely shared. Jonathan Kuniholm, one of the founders and forces behind OPP, describes this purpose in an interview with *Scientific American:*[16]

> "The reality . . . is that there's no traditional economic incentive to do work and make improvements on prosthetics. That doesn't mean that nobody cares, but most people don't have the money or know-how to magnify whatever efforts or improvements they make. I think we can generate

far more societal benefit if we give away information than if we commer-cialized and sold the ideas. Our goal is to create a way to share these efforts and improvements with anyone who needs them."

The project is clearly a successful example of the power of collaborative communities, though the connection with financially viable businesses is yet to be proven. In a real sense, the lack of IP protection makes it difficult for small companies to take the risk in manufacturing a design that anyone can freely reproduce. Nevertheless, there is a real social good, and a decidedly higher rate of innovation taking place than had in the past.

Wikipedia

Whatever you think about Wikipedia's reliability compared with "well-researched" sources, you should pay very close attention to the community model. In this regard, the success of the site is stunning. Wikipedia.org is well within the ten most popular Web sites, according to Alexa,[18] garnering around 250 million unique visitors each month.

This is the territory of significant commercial concerns, such as Yahoo!, Google, Microsoft, and Facebook. Yet fewer than a couple dozen people are employed by the Wikimedia Foundation, which operates on a shoestring budget. The secret, of course, is that there is an enormous amplifier in the Wikimedia contribution community, which the Wikimedia Foundation pegs at about 100,000 people, with a few percent being the most active. We think the licensing terms on the content have played a key role. As the Wikipedia entry on "Wikipedia" describes:[19]

> *All text in Wikipedia is covered by [the] GNU Free Documentation License (GFDL), a copyleft license permitting the redistribution, creation of deriv-ative works, and commercial use of content while authors retain copyright of their work. The position that Wikipedia is merely a hosting service has been successfully used as a defense in court. Wikipedia has been working on the switch to Creative Commons licenses because the GFDL, initially designed for software manuals, is not suitable for online reference works and because the two licenses are currently incompatible.*

What is interesting to watch with this community is the way that collec-tive content is continuously refined. All page entries are subject to "vandal-ism" whereby a rogue contributor provides deliberately misleading or biased information. Interestingly, many of these contributions are corrected (by the community) in a matter of a few minutes, or perhaps a handful of hours.

We think that collaborative content development and maintenance is a key aspect of healthy communities, independent of discipline. The basic model popularized by Wikipedia looks to be a sound one and worthy of emulation.

OpenSPARC

If you peer over the shoulder of many digital hardware designers, you'd swear they were software engineers: They are in front of text editors writing code that looks like C. That code, written in high-level concurrent textual languages called Hardware Description Languages (HDLs),[20] describes the intended behavior of logical circuits. Many simulation and verification tools work on the high-level code, as do logic synthesis programs that essentially compile the code into digital hardware.

So, the concept of "open source hardware" is really a simple one: Develop the HDL code under an open source software license and try to foster open communities. That's exactly what we did at Sun with the OpenSPARC project, but there are many other examples as well—Opencores.org being a leading one.

With OpenSPARC, we released the HDL code (contained in thousands of files) under a GPLv2 license. We chose that license because we wanted to create a strong sense of sharing in the community. Not only does the license require put-back, but we are hoping that the viral aspect will encourage other "adjacent" contributions to be made as well (e.g., a new way of doing networking). The GPL also creates a commercial opportunity to sell the source code under license to those companies that, for competitive reasons, *don't* want to release any changes, improvements, or extensions they make to the design.

It's worth noting that, for high-performance digital systems, there is a big gap between the HDL code and a working chip. Certainly, automatic synthesis of entire chips is possible, but it's a lot of work involving a lot of engineering labor to get to an implementation that is competitive. So, there are natural barriers in this discipline to prevent someone from simply copying the physical design.

So far, we view the OpenSPARC program as enormously successful in building a global innovation community outside of Sun. There have been thousands of downloads of the design files and many universities have incorporated OpenSPARC into their curricula. At a minimum, we think this program illustrates how other engineering communities can be built around an open model.

Building an Open Source Community: Practical Advice from a Pro

"If you build it, they will come," whispers a voice from the cornfields in the movie *Field of Dreams*.

Many software engineers seem to hear the same voice. Having created some interesting code, they want to broaden its exposure. They want to entice others to use, amplify, propagate, and promote it. So, they create an open source community. And they learn the hard way that building a thriving community of users and creators can be more challenging and less likely to succeed than, say, building a professional-grade baseball field on a farm in Iowa.

We asked Kaj Arnö, who was instrumental in building the MySQL community (which we consider to be one of the most important communities in the annals of open source software), to offer some advice about building open source communities. Here's a summary of what he had to say.

Q: What makes a good candidate for an open source project?

A: You need to have a very cool technology; it needs to solve a core problem that many people share; and it needs to solve the problem better than anything else that's available.

Let's break that down, starting with the "cool" factor. A good test is, do people's jaws drop when they first see it or use it? Do their eyes open wide? If not, it may be simply an incremental improvement to something that's already out there, or it may not have broad appeal, or it may not be clear what the value is. In any of those cases, it's missing the spark that will ignite the interest in a community.

Now, it must also solve a common need in a novel way. You can bet that if you're working on a solution to a big problem, then someone else is too. Is your solution clearly better? If the advantage isn't obvious, it may be difficult to build a community around it.

Q: Assuming your software passes the "jaw dropping" test, why build a community? Why not continue to develop it yourself, or internally within your company?

A: A community can accelerate the propagation of your software dramatically. And I'm not talking only about using a community to recruit code contributors. Even in the MySQL community, not many people actually write code for us.

First of all, you don't need to explain all the basics to the people who join your community. In the case of MySQL, you don't need to tell them what a database is for. They know; they get it; they don't need time to come up to speed. They're ready to participate right now.

Second, the community is great for quality assurance. They want to help you test things, find bugs and list them, and so on. It's a very focused effort.

Third, as the community gains momentum it can deliver access to people who know a whole lot about the type of software you're working on. It becomes a highly concentrated source of a very specific expertise.

Q: What's the "right" way to go about building an open source community, and what mistakes do you see others make?

A: What you're trying to build is an architecture of participation. And you can't do that by blogging and giving away T-shirts. The key is not to "push" the community yourself or internally, but to keep the focus on building momentum externally.

Yes, it's a chicken-and-egg problem: You can't grow the community if you don't promote it, but the more actively you promote it the greater the risk of alienating people. A good starting point is to identify a set of core users who are highly informed and skilled, and spread the word. If your software has clear value, those people will quickly become evangelists.

At the same time, there are several things you can do to grease the wheels. First, do everything you can to make your software easy to get and easy to use. Make sure visitors to the community Web site can find the software, download it, and configure it easily. If they like what they're using, they'll see the value and find ways to contribute.

On a related note, make sure your documentation is good. If you can offer excellent documentation you will greatly increase the odds that people will want to work with your software. Documentation writers are a special breed of people—find good ones and keep them happy!

The design decisions you make relating to infrastructure are key. If you explicitly tie your software with a certain processor architecture or one vendor's systems, for example, you will limit the adoption rate of your software. My advice is to avoid tight integration with specific hardware whenever possible.

It's also important to have a good Web infrastructure for your project. As I mentioned, make it simple to find, download, and configure the software. But also make it easy to interact with others on the site. As the community grows, add forums, general usage discussions, and blog aggregations. Get a channel on Freenode. Start offering training and online support.

Q: But aren't some open source participants turned off by a glitzy Web site?

A: I think "slickness" can easily backfire because people associate it with a lack of substance. But there's nothing wrong with having an attractive Web site that's easy to navigate! There's a big difference between being efficiently functional and being glitzy. In the early stages of MySQL, there was absolutely no marketing on our Web site. Everything was technical; everything was written in a vendor-neutral way. I think people appreciated that, and we still try to focus on substance, objectivity, and honesty over style.

Q: What should you measure in assessing the community's growth?

A: Most organizations have the desire to measure and quantify the performance of everything, including community growth—but measuring can actually get in the way of growth if it's not done correctly or if you measure the wrong things. Always keep the focus on finding out how much users like what they're using. Two good measures are software downloads and documentation downloads. If they're using the software, if they're looking at the documentation, those are very good indicators that they like it. Other measurements, such as Yahoo! hits or click-throughs resulting from Google searches, can be misleading. You don't know if it's good word of mouth from other community members or clever search engine optimization techniques.

Other positive indicators could include bug activity, such as the number of filed bugs, verified bugs, and fixed bugs; the number of emails sent to the product email list—by both employees and nonemployees; blog and Wiki activity; and user popularity metrics such as the number of subscribers to your newsletter or email lists, downloads of the product on external sources such as SourceForge, [the] number of registered contributors, active users, signed contributor license agreements, and so on.

Q: What are some of the licensing issues you need to consider when building an open source community?

A: The choice of a license is just a step, because building a successful community of users and creators is the real prize. But the license decision is important because licenses are constitutions for communities. There's no general guidance I can offer—the choice will depend on the nature of your software and possibly your company's policies regarding IP ownership. Licenses such as BSD and MIT are very permissive; the GPL licenses are more restrictive and often require put-back and patent grants; and others such as Apache, CDDL, and Mozilla fall in between. Whatever license mechanism you choose, be clear about it and make sure your community knows exactly what it means for them.

Q: How can the founder of the open source community maintain control without being overly controlling?

A: Stay open and honest. In every community there will be participants who don't like the direction the project is taking, who have complaints about other participants, or who propose things that aren't in keeping with the community's values or are just bad ideas. Tell people what you think and you'll maintain control by maintaining respect. Also, don't feel compelled to be right all the time. Occasionally, I'll see a founder try to rule authoritatively, constantly asserting his or her superior knowledge. That approach usually leads to forking. If somebody's not happy, let them fork; usually if the fork is based on some personal motivation, it won't succeed. But in general, it's better to try to accommodate diverse viewpoints and encourage broader experimentation.

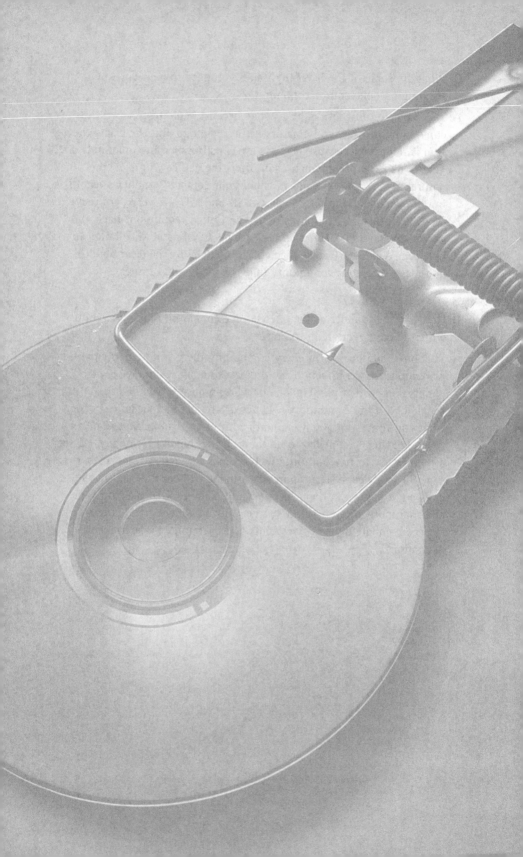

15

Protecting Digital Rights

The volume of digital content coursing through corporate and consumer networks is on upward spiral at least as steep as Moore's law. The result is that the total amount of content available online has exploded—and so have the number, variety, and creativity of the exploits designed to circumvent protection mechanisms for digital content. It's a perpetual arms race between black-hat hackers and white-hat engineers—and neither side is ever ahead for long.

What can a Citizen Engineer do about it? Plenty. Here are a few insights, ideas, and considerations that might help you protect consumers—and yourself—from unauthorized access to data and content.

Digital Rights Management

One reason innovation has been able to flourish is because Internet technologies enable the rapid, widespread, and often anonymous flow of information. Combine that free flow with advances in digital media—photography, video, music—and you have an amazing opportunity for wide-scale experimentation and creative expression.

Two decades ago, home computers brought us a revolution called desktop publishing. Now home users have the tools to create professional-quality movies and music—and a way to share them with others.

More recently, though, the Internet has become a place of conflict and contention when it comes to digital media. The questions are difficult: Since the network makes it easy for people to copy and transmit protected material, how can we make sure artists are compensated for their creative work?

And just as important, how can we do so without quashing experimentation and innovation?

Artists should be compensated. There's no question about that. But in our rush to defend their rights, let's not overlook the need to encourage experimentation. As Citizen Engineers, we believe public policy should encourage innovation and free speech. It should, as always, seek to balance the rights of individuals with the greatest public good.

In the words of Lawrence Lessig, "The Internet has created a great fear among the content industry that they're going to lose their whole industry if they don't learn how to, and get more power to, exercise perfect control over content. And so, what they're doing is building both technology and a legal infrastructure to give them much more control than they've ever had in the past."

But one of the great values of the Internet is that it has become a forum for borrowing, mixing, developing, and tinkering. After all, in both science and art, innovators build on each other's work.

So, as we look at the developing discipline of digital rights management, or DRM, we need to respect the experimental, standing-on-the-shoulders-of-giants aspects of the Internet. It's up to all of us who see ourselves as Citizen Engineers to advocate policies and design technologies that respect the legitimate needs and current rights of honest users.

While the Internet certainly makes managing the rights for movies and music more complex, we believe that it is a more sound economic and social policy to foster the architectural, business, political, and public freedoms that have enabled the Internet to be a place of innovation than it is to overly restrict the flow of digital information in an effort to meticulously account for every instance of the use of content.

We think the rights of content creators can be balanced with the common public interest to foster vibrant innovation. To that end, we'd like to suggest that the following principles be applied to digital rights management.

- Innovation flourishes through openness.

- All creators are users and many users are creators.

- Content creators and copyright holders should be compensated fairly.

- Respect for users' privacy is essential.

- Code (both laws and technology) should encourage innovation.

Some content owners are pressing for DRM systems that would fully control the users' access to content, systems with user tracking that would limit access to copyrighted material. We instead prefer an "optimistic" model whose fundamental credo is trust the customer. Excessive limitation restricts not only the rights of consumers but also their potential, because such solutions strongly interfere with the creation of derivative works and fair use of copyrighted content.

In an ideal world, solutions should encourage information flow, including the capability for creating derivative works. While we recognize that there will always be "leakage" and illegal behavior, we think it's better to provide auditing and accounting paths that respect the privacy of honest users and permit copying, manipulation, and playback.

As Lessig puts it: "The counterintuitive part about intellectual property is that there can be too much of a good thing; that if you exercise too much control, or if a law grants creators or innovators too much control, that can actually stifle new innovation. So, the really hard problem for policymakers, and also for companies dealing in this context, is to strike the right balance between the control that's necessary to produce the profits that will support new innovation and the access or freedom of others in that space, so they can build on top of that innovation and make it worth much more."

Systems that encourage the user to play with digital material, to experiment, to build and create, will be a win not only for consumers but also for content producers.

Is "Open DRM" an Oxymoron?

Digital rights management is all about control, restrictions, and authorizations, seemingly at odds with the spirit of community and openness that is the hallmark of the Citizen Engineer.

So, why have Sun and other organizations invested so much time and talent in developing a standards-based, royalty-free architecture for building interoperable DRM implementations?

Because an open, interoperable DRM architecture can help to solve problems that proprietary DRM technology created. And because it can open up new opportunities for companies that want to expand their digital content offerings without compromising control and protection of their content.

You don't have to look very hard to see the limitations of proprietary DRM technologies. What do you do when you want to watch a movie you've downloaded legally, but at a friend's house instead of at your house? Your friend has a different movie player than you do. The DRM system isn't smart enough to recognize that this would be legitimate access. Today's DRM model is like the old AOL model: As long as you buy into the AOL world and stay in the AOL world, you're fine. But if you venture outside, trouble is lurking.

A better approach is to create an open, standards-based DRM architecture and to share it—royalty free—with the development community so that they can innovate and add value without technological encumbrances or prohibitive licensing costs.

"In a world where DRM has become ubiquitous, we need to ensure that the ecology for creativity is bolstered, not stifled, by technology," says Lessig.[1] "We applaud efforts to rally the community around the development of open source, royalty-free DRM standards that support 'fair use' and that don't block the development of Creative Commons ideals."

The concept of interoperability is at the heart of the matter. For there to be a diverse and robust economy supporting the sale and use of creative content, DRM systems must be interoperable, and this interoperability must encompass all necessary functionalities. It is critical that not just fair use rights but also user rights and rights on contract termination be present in all DRM instances, and that user interests be considered and fully included.

We believe that DRM should be a solution *only when necessary*. DRM should never restrict the user's ability to utilize the content in legally permissible ways. Without full interoperability, users find it difficult and even impossible to listen to and view content they have legitimately purchased. Instead, users are locked in; and we believe such limitations on the use of legitimately owned creative content represent an unacceptable abridgment of consumer rights. As important as the issues of control and appropriate compensation for content owners are, interoperability is even more critical for all concerned.

Sun has a lot of experience in this arena; it is, after all, the company that believes "the network is the computer." Sun's work in this area is fundamentally based on open standards, and increasingly on open source. Sun has already launched an open source project in DRM interoperability, Project DReaM, whose source code has been released as part of Sun's Open Media Commons initiative.[2]

Project DReaM[3] provides the infrastructure to develop an open DRM solution for consumers, content owners, network operators, and device manufacturers that strives to chart a royalty-free approach. This is one solution to the interoperable DRM problem; there are undoubtedly others.

Fair Use and Other Concepts for Reducing Restrictions

It is also important to recognize that Internet users are active content creators, not just passive consumers of content. User-created content—blogs, videos, music, and mash-ups—is one of the most vibrant areas of creation on the Internet, and embracing this key shift should help to keep DRM inter-operability in its rightful and constructive place in the online world.

A vital aspect of enabling creative content is "fair use," the American legal concept that in certain instances creative content can be used without the content owner's permission. Some members of the European Union share some similar ideas under "fair dealing." But while the law may protect fair dealing, DRM technology does not necessarily do so. Placing restrictions on the use of creative content is not good public policy, and any policy on DRM should include support for legally permitted uses of content.

An important component of support for user-created content is Creative Commons, which, as noted throughout this part of the book, provides licenses enforcing some restrictions on the use of content (e.g., Attribution Required, No Commercial Reuse) while encouraging the sharing of content. These licenses have seen wide international adoption, with licenses available in 53 nations. We consider Creative Commons methodologies a useful best practice foundation.

Another important aspect is the handling of the rights to the content on termination of the contract that the DRM system may be attempting to implement. DRM inherently "quantizes" rights, removing the nuanced ability to interpret rights that has traditionally been available. For the market to be able to work with the base of cultural material being generated today, it is important that DRM systems include the ability to handle "end of life" for services and systems without either removing the rights granted under the contract with the supplier or blocking the ability to exercise statutory freedoms.

We recommend a forward-moving strategy based on four basic foundations.

- **User-centered principles:** This entails recognizing and further enabling a user-centered right to use content on multiple clients, archives, and devices; a user-centered right to assert license and ownership as expressed in techniques such as Creative Commons; and a user-centered right to know license and usage restrictions (and associated costs) on both content and the media technologies needed to use the content.

- **Adequate stakeholder scope:** Implementing meaningful DRM interoperability requires consideration of not only multiple devices, services, and archives, but also multiple networks and associated business models. We believe it is important to consider both the technical and business model similarities between DRM and conditional access, and to consider that similar obstacles have limited interoperability in both domains.

- **Acceptably inclusive methodology:** We believe that royalty-free (while reserving defensive rights), open source, and ex ante standardization processes for DRM interoperability are the best, fairest, fastest, most inclusive, and most certain route to meaningful DRM interoperability.

• **Recognition of cultural priorities:** We believe that the longer-term needs for access to cultural artifacts require prioritization alongside commercial issues. This would mean, for example, ensuring that "locks" on rights can be removed at contract termination, or when a service provider leaves the market, so that interoperability can still be achieved even in the absence of the original provider.

DRM systems without full interoperability will likely lead to markets for devices and content being severely hampered, which, in turn, could easily lead to a monopoly situation and greater restrictions in consumers' ability to use content they own. Neither of these situations will serve culture, business, or democracy particularly well.

The raucous debates about DRM continue among computer industry executives, Hollywood moguls, intellectual property lawyers, members of standards organizations, and content owners of all types. Ultimately, all of those issues are for the marketplace to decide. But as long as DRM remains part of the equation, as long as DRM technology is instrumental in the management of content distribution, we need an open, interoperable DRM solution, not a proprietary product.

Part III Summary, and What's Next

This part of the book gave you but a small taste of the opportunity—and the complexity—presented by intellectual property law. While we have only scratched the surface of this topic, we hope it's enough to get you to start seeing IP mechanisms as a powerful lever, not just as a potential minefield for your project or your career. Learn to turn IP law to your advantage and you'll find that it will enable your ideas to thrive, to evolve, to spark new innovations within your teams and communities, and ultimately to succeed in the marketplace. If you want your stuff to win, you've got to know your stuff when it comes to IP.

The next part of the book takes us beyond the realm of practical information and addresses a fundamental question: How does the "ideal" of the Citizen Engineer become a reality in practice? We'll provide advice and examples from around the world.

PART IV
Bringing It to Life

By this point in the book you should have plenty of food for thought about the opportunities and the challenges that await you in the era of eco-responsible, techno-responsible engineering. But how do you put all of this into practice? We've talked about how engineers can use the mechanisms of IP law to propagate their ideas; now it's time to examine how societal mechanisms such as schools and businesses can help to propagate Citizen Engineers. Here are some questions we'll consider.

- Should the core curriculum in engineering schools be broadened?

- How can recent engineering school graduates and newly hired engineers make an impact right now?

- What kinds of projects are real-world Citizen Engineers undertaking today?

This part of the book addresses those broad issues and takes a closer look at some of the ways engineers, engineering schools, and organizations that employ engineers are responding to the new realities and requirements of the new era.

16

Education of the Citizen Engineer

Where and how do you as an engineer strive to become a Citizen Engineer? Right now there is only one answer: through your own initiative. Although many schools are developing and evolving programs in engineering ethics and the relationship between engineering and society, there is no formally accredited curriculum at a university, no on-the-job training program at a corporation, and no comprehensive seminar or online resource.

In the meantime, there are things you can do. We'll assume you always are going to strive for excellence in your core engineering discipline; that's one of the hallmarks of a great engineer. To become more of a Citizen Engineer, we urge you to do the following.

- **Learn** the relationships between what you do and the broader social interests of the environment, safety and trust, security and privacy, choice, and competition.

- **Understand** the law and public policy. If you complain about the general level of ignorance others have about science and engineering, turn it around. What's your level of understanding about the legal and political systems?

- **Participate** in public dialogs regarding these topics. As an engineer, you bring skills and gifts to your local and national communities—from analytic reasoning skills to your constructive art. Think of these communities as your customer; listen, engage, and serve.

- Act on what you know and believe. Help to build innovation commons and vibrant communities that transcend your company and your country: Participate, contribute, grow, and help them.

We are unapologetically idealistic: That is a very full list. But if we are to lift our profession and truly help to lead humanity through this century—the century of engineering—we have to imagine the possibilities and work toward them.

At the core is education—expanding the very notion of what an engineer is and growing collectively who we are. Here's a brief synopsis of what schools are doing to facilitate the education of the Citizen Engineer, along with advice from some of the people we've spoken with in preparing this book.

With apologies to our global readers, we'll focus primarily on engineering education in the United States. While the primary education issues—especially the state of math and science education—may currently be unique to the United States, the broader analysis regarding university curricula is not. The education programs of U.S. engineering institutions continue to influence programs at universities worldwide.

Updating Engineering Curricula

> "The education system needs revamping. You're teaching the wrong stuff, which is why I'm on your case all the time. You're still teaching the system that is destroying the biosphere, and teaching the teachers to perpetuate it."[1]
> —Ray Anderson, chair, Interface, Inc.

That assessment was made several years ago by one of the leading advocates for industrial ecology and sustainability. Does it still hold true today? Yes and no.

Clearly, something about the way science and engineering are taught in the United States needs revamping. Over a 27-year period, from 1975 to 2002, the percentage of 24-year-olds in the United States who earned first Science, Technology, Engineering, and Mathematics (STEM) degrees increased by 43%; during this same period, that number quadrupled, on average, in Taiwan, South Korea, France, Spain, Mexico, and China.[2]

And according to a recent article from the National Academy of Engineering, "Overall, the number of engineering B.S. degrees earned by U.S. students peaked in 1985, steadily declined through 1992, and then came to rest on a decade-long plateau. The number began to climb again in 2002, but is still lower than it was in the mid-1980s. Coupled with a dramatic increase in retirements expected in the next two decades, these numbers signal a national imperative that we attract more—and different—U.S. students to the engineering fold."[3]

If U.S. students are turning away from the field of engineering, it is incumbent upon all of us to ask why. One crowd says U.S. students don't think they can compete with the huge influx of engineering talent from Europe and Asia; another crowd says the bursting of the dot-com bubble has removed incentive because engineers can't make the quick millions anymore. We don't believe either of those things. We've seen no change in the number of very talented, technically savvy engineers coming out of U.S. colleges. They seem to get smarter every year, largely because of things such as open source software, which increase the volume of knowledge available to them.

Clearly, there are many reasons why the ranks of U.S. engineering students aren't growing as quickly as we'd like. One of those reasons, in our opinion, is that U.S. engineering schools are not fully tapping into the energy that's building around environmental and techno responsibility. Students really do want to change the world, and are very focused on making a difference in the critical global issues of ecosystem sustainability, health, education, economic opportunity, and human rights. Engineering and technology can have profoundly positive effects on all of these, but our schools have to be deliberate in making these connections, or else the best students will be attracted to other avenues.

To be fair, a growing number of engineering schools are beginning to recognize the growing importance of sociological, environmental, and intellectual property issues and are mixing these topics into the traditional curriculum. Better yet, some schools are introducing an interdisciplinary approach and creating new courses that attempt to broaden the field of study—and the perspective—of engineering students.

For example, at the MIT Sloan School of Management, Professor Steven Eppinger (currently deputy dean) has created an interdisciplinary product development course in which graduate students from engineering, management, and industrial design programs collaborate to develop new products.

"An interdisciplinary approach is important because that's what it takes to develop successful products," says Professor Eppinger. "In today's companies, innovation processes are collaborative and team-based. Critical inputs to product development come from many directions—engineering, sales, marketing, finance, even legal. Yet universities have traditionally taught in a stovepipe manner. We teach industrial designers separately from engineers, separately from business students, separately from lawyers, and we leave it up to the students to make the connections between the disciplines. For this new generation of young engineers, those interconnections are vitally important. If you're not content with designing just any old products but you want to design products that really make a difference, engineers need to know something about business, about law, about the environment, so they will ask the right questions and seek inputs from other experts."

Many other schools are now preparing interdisciplinary courses or launching initiatives specifically focused on environmental engineering and sustainability. Here are some examples.

- Michigan State University offers a master of arts degree in environmental design, bringing a multidisciplinary approach to professional development, including acquisition of in-depth knowledge in the area of environmental design theory; development of problem-solving skills within an interdisciplinary professional context; development of technological expertise and knowledge base in a selected area of environmental design; and advanced ability in graphic, written, and oral communication skills.

- Cornell University now offers a minor in environmental engineering, encouraging engineering students "to learn about the scientific, engineering, and economic foundations of environmental engineering so that they are better able to address environmental management issues."[4]

- At Kettering University, a multidisciplinary engineering elective course employs proven pedagogical methods and tools that enable students to incorporate environmental and economic concerns into technical designs.[5]

- Virginia Tech offers a bioprocess engineering specialization, which combines knowledge of biological, chemical, and engineering principles to produce sustainable and environmentally responsible food, fuels, pharmaceuticals, plastics, construction materials, and other products from biological materials.[6]

- Introductory engineering courses at Michigan Tech now emphasize communication skills as a core element of engineering problem solving.[7]

- The College of Engineering and the Jackson School of Geosciences at the University of Texas jointly offer a degree program designed to teach students the geological and engineering principles needed to solve resource development and environmental problems.[8]

These and many other similar efforts are a step in the right direction. However, the shift to more broad-based, interdisciplinary, and/or environmentally responsible curricula has been painfully slow at many engineering schools.

"I've seen a trend toward more social consciousness both in process design engineering as well as in manufacturing processes, and that is beginning to be reflected in the curriculum here at Wisconsin and at other institutions," says Professor Harold Steudel of the Department of Industrial and Systems Engineering at the University of Wisconsin-Madison. "For example, our Department of Professional Development has introduced new courses on environmental management and sustainability. But it can be a slow and difficult process—both logistically and politically—to redesign [curricula] at major institutions."

Typically, change comes to the curricula at public universities through the efforts of a tenured professor or a faculty member with the energy and enthusiasm to push through new courses. And all too often the inertia outweighs the enthusiasm. So, what can an aspiring Citizen Engineer do? Here's some guidance.

Advice for Engineering Students

"You can learn to program and you can get a job fixing bugs . . . but if you want to do something that makes a difference, you have to learn to think across boundaries, to understand business, customers, law, public policy. . ."
—*Mike Shapiro, Distinguished Engineer, Sun Microsystems*

It can be a tough balancing act. If you broaden your areas of study while you're in engineering school, potential employers may consider you to be too unfocused. A triple major, for example, may actually be counterproductive. On the other hand, if you narrow your field of study, you may limit your growth—personally and professionally.

Good schools are starting to ensure that engineering students get deeper knowledge of how things work across traditional boundaries—for example, making sure that software engineers understand how a microprocessor works, how to write a compiler, how to write a program on top of that, and so on. Another dimension of this increased breadth is helping engineers understand how to analyze a market, how to explain the customer benefit of a new innovation, or what it means to have a great idea that has no channel to customers.

Today's responsible civil engineer must be aware not only of technical design issues but also of more efficient and "greener" materials, socially and environmentally responsible construction methods, and the need to collaborate closely with architects to achieve the best combination of art and functionality.
—*Ricardo Davila, MIT '06*

Here is one strategy that many people advocate for would-be Citizen Engineers: While in school, focus on getting that first job. Be technically savvy in your area of specialty, then broaden your base of knowledge and skill sets as your career evolves. Here's the problem with that advice: Narrowing down is very counterproductive in the long run. All the really interesting innovation occurs across boundaries, not within a specialty. For students, it's more important to get a deeper appreciation of interrelationships and to learn how to understand something that's completely new.

Our advice is to be as broad as you possibly can, especially as an undergraduate. Be sure to take courses in economics, business, political science, and law. Focus on how these areas help you *reason and think* rather than simply catalog knowledge. Legal reasoning, for example, is fundamentally different from what you are used to. Of course, logic does apply, but for many legal systems it is the history of the field in terms of case law, so-called *legal precedent,* that forms the underlying axioms. In these systems (e.g., in the United States), you don't argue what ought to be true or just; you argue how something relates to the case law. And for all systems, process is paramount— *how* something is decided may be more important than *what* was decided.

"Learn how to learn" may be hoary, but it's great advice. Learn how other people learn and reason too. Your best bet in influencing what happens in some other sphere, such as law or public policy, is to speak their language, rather than expecting them to speak yours. You will also be amazed how, later in life, you draw as much on your education in areas such as business, law, and ethics as you do on your core courses.

Advice for Engineering New Hires

"Don't just answer the question what are we building; ask what could we build?"
—*Sheueling Chang,*[9] *Distinguished Engineer*

One of the key concerns of many newly hired engineers is that they'll have very limited input into any social or environmental considerations of the projects they're working on. At the surface level, this often seems to be the case. If you're hired to optimize the firmware on a new board design for a mobile handset, management is not looking to you for guidance about the company's take-back policies. You have specific deadlines and deliverables. So, when and how do your values as a Citizen Engineer enter the picture?

In the words of Sun engineer Mike Shapiro, "If you're passionate about anything in your life, you're heading in the right direction. If you're passionate, you have to ask yourself: Am I willing to really learn my craft and learn how to be a leader? If you are, you will notice two things: You will be able to engage in any activity or project and learn something by doing it, and you will become a better craftsman. A lot of people think what they're doing is unimportant. That's sometimes because they're not passionate; it's more often because they are passionate but they don't understand that what they're doing now will enable them to become a better craftsman and a better leader. If you're committed to being a craftsman, study the work of others. You'll learn how to be a leader, and you'll develop influence. No project is too trivial or simple to learn something valuable from."

Michael Falk, general counsel of the Wisconsin Alumni Research Foundation (WARF), adds the following: "As an engineer you have a unique set of abilities that so few of us have—the ability to manipulate what the real world looks like. Whatever your role is, expand the paradigm of the question you've been asked. If you allow it to be pushed into a narrow arena you cheat yourself and you cheat everyone else, and you limit your role. Being an engineer isn't just about solving discrete problems; it's about answering problems in unconventional ways. That's the kind of creative work people want engineers to do. There's always an opportunity to do it."

17

Citizen Engineers in Action

> "We package engineers as problem solvers rather than creators and innovators
> who address the grand challenges of our time—environmental contamination,
> world hunger, energy dependence, and the spread of disease . . .
> How did we let this happen?"
> —*Jacquelyn F. Sullivan,[1] co-director of the Integrated Teaching and Learning*
> *Program at the University of Colorado at Boulder*

Around the world, Citizen Engineers are making a real difference in improving the quality of life. Some are working in the companies you pass by every day, making a difference in the products that we use in our daily routines. Others are applying their passion and expertise to solving fundamental problems that people face. As a conclusion to this book we thought we'd highlight a few inspiring examples of the kinds of things real-world Citizen Engineers are working on today.

Engineers Without Borders (EWB), a nonprofit humanitarian organization, is partnering with developing communities worldwide in order to improve their quality of life. This partnership focuses on the implementation of sustainable engineering projects, while involving and training internationally responsible engineers and engineering students. Here are just a few of their recent projects.

- In Bulandshahar, Uttar Pradesh, the student-teacher duo of Niruttam Kumar Singh and Harvansh Yadav have made a cow dung battery that lights up electric bulbs, charges mobile phones, and brings alive radios.[2]

- Undergraduate engineering students are currently building a bridge across a gorge in a small town in Nicaragua. The students have surveyed the entire project site and are now in the process of designing a bridge to span the gorge and allow for pedestrian travel during the rainy season.[3]

- Thousands of residents of rural villages in India are receiving quality eye care thanks to a collaborative effort between an Indian hospital network and the researchers at the University of California, Berkeley, and at Intel Corporation who have developed a new technology for low-cost rural connectivity.[4]

- Engineers at PlayPumps International designed the PlayPump[5] water system, which provides easy access to clean drinking water, brings joy to children, and leads to improvements in health, education, gender equality, and economic development. Installed near schools, the PlayPump system doubles as a water pump and a merry-go-round. It also provides a way to reach rural and peri-urban communities with potentially life-saving public health messages.

In Panama, students and researchers are using small wireless sensors to help answer big environmental questions. Warren Wilson College and CREA, a nonprofit organization in Panama, are implementing a geographic information system (GIS) and wireless sensor network on the 1,000-acre Cocobolo Nature Reserve in Panama. Tiny Sun SPOT sensors[6] will provide an inexpensive, easy-to-program platform for monitoring all kinds of things: the impact of deforestation on an ecosystem, plant and insect activity in a rainforest canopy that's 60 feet off the ground, or small changes in local atmospheric conditions that reveal broader meteorological patterns. "This network will allow students to ask big questions and get meaningful answers," says Warren Wilson College Geography Professor David Abernathy, who is overseeing the implementation of the sensor network. "We're extremely excited about the possibilities for our research."

At Rice University, graduate students are using nanotechnology and biotechnology to create high-performance and cost-effective water treatment systems and create the information needed to ensure that emerging technologies evolve in an environmentally responsible and sustainable manner.

Engineers at Tesla Motors will have a profound impact on the environment—whether or not their start-up company succeeds in the marketplace. By proving that a high-performance electric car with zero exhaust is now technologically feasible, Tesla engineers have already radically altered consumer attitudes about electric vehicles and accelerated industry-wide development of new energy-efficient technologies.

Engineers at Global Research Technologies (GRT), a technology research and development company, and Klaus Lackner from Columbia University have demonstrated a new technology that captures carbon from the air. In the "air extraction" prototype, sorbents capture carbon dioxide molecules from free-flowing air and release those molecules as a pure stream of carbon dioxide for sequestration.

This new technique has met a wide range of performance standards in the GRT research facility. "This is an exciting step toward making carbon capture and sequestration a viable technology," said Lackner in an interview with The Earth Institute at Columbia University. "I have long believed science and industry have the technological capability to design systems that will capture greenhouse gases and allow us to transition to energies of the future over the long term."[7]

And you don't have to be an international conglomerate to practice the lifecycle approach to engineering. As showcased in an issue of *Newsweek* magazine, a small-scale project in Brazil shows how collaborative engineering can create an environmentally responsible business that also benefits the environment.[8] José Roberto Fonseca, an engineer and environmentalist, found an opportunity for farmers to grow their way out of poverty. He devised a scheme for using solar power in a desolate, semidesert area of Brazil to irrigate suspended gardens of red, orange, and yellow hot peppers, which could then be chopped, bottled, and exported as gourmet vinaigrette.

Fonseca's solution was based on hydroponics. The pepper plants are grown in water laced with nutrients on a wooden trellis crisscrossed with ultrathin irrigation tubes. At first, his team drew well water and filtered away the salt using a solar-powered desalinator. Now the community taps a natural spring and lets gravity bring the water to the plants. A bank of photovoltaic (PVC) panels powers pumps that keep the water flowing. A "daisy chain" of inventors and entrepreneurs are involved in the production process. An agronomist and an engineer designed the hydroponic gardens; a nutritionist taught villagers the secrets of making spices and condiments; an economist worked up a business plan; and Fonseca built a distribution network with start-up money from international benefactors. "He thought it through, from the soil down to the dinner table," says John Ryan, head of the Virginia-based Institute for Environmental Development.

Today 11 family businesses in Baixas are making a good part of their income from the peppers, and economic prosperity has come to one of the poorest places on Earth—without harming the environment.

In another example, a partnership between Audi and UC Riverside (along with UC Berkeley) has resulted in a project called "Clean Air, a Viable Planet," announced in fall 2007 at the Los Angeles Auto Show. The goal of this project is to reduce CO_2 emissions by allowing drivers to determine the greenest route possible in current traffic conditions. The theory is that any vehicle, regardless of its fuel economy rating, will use less fuel getting from point A to point B if it can cruise at a constant speed rather than if it is constantly speeding up, slowing down, and idling in traffic. "Our goal is to be part of a real solution to the constant dilemma commuters face: What is the best way to get there?" said Matt Barth, professor of electrical engineering and director of the College of Engineering Center for Environmental Research and Technology (CE-CERT), in an interview published on the UC Riverside Newsroom Web site.[9] "Sometimes the best way to get there is the one that causes the least damage to the planet."

Appendix

Lifecycle Phase Checklists

These checklists summarize the important items that can affect the legal standing, business results, and environmental impact of your product.

If it hasn't been obvious already, you'll notice that most of these items need to be addressed at design time. Even though they may relate to later phases of the product lifecycle, after the product or service has been designed it is usually too late for these items to have a serious impact on the overall design. However, there may be ongoing opportunities to switch parts or vendors, redesign packaging, change logistics strategies, and so on, and make ongoing improvements to a shipping product.

Another key point is that your product will be subject to the laws of the countries, states, provinces, and other jurisdictions where you intend to sell it. It is also important to understand how your product will be characterized in each jurisdiction, so you can know which laws will apply.

Finally, we assume that you've been able to do a high-level lifecycle analysis of the energy usage, greenhouse gas (GHG) emissions, and water usage of your product so that you can use it to guide your decisions.

The "Make" Phase

As we've described, our focus on the "Make" phase leads us to think about how products are manufactured and delivered to customers; in general, we don't need to worry about this phase for services. For some products it may be important to look at the impact of design and testing as well, but for higher-volume products this is often a minor impact. In this phase, we're often focused on chemicals that make up a product, energy to make a product, packaging, and logistics. Here's a checklist of key items for the "Make" phase.

"MAKE" PHASE CHECKLIST

Legal Implications

____Make sure you understand and can meet all relevant chemical and substance laws that will apply to your product.

Business Opportunities and Threats

____Know whether you are using any materials that may involve destructive or socially suspect extractive processes.

____Minimize packaging and make it easy for your customers to responsibly dispose of it.

____Understand your use of materials and natural resources, and understand the cost impacts of historical market prices or future pricing scenarios.

Major Impacts

____Use your lifecycle analysis to see whether there are major impacts during the "Make" phase for which you'll need to create a plan over time, including energy usage, GHG emissions, and water usage (you should minimize these as part of your design, but also put in place a plan to focus on and reduce these over time).

____Understand any hazardous chemicals or emissions that will result from the manufacture of your product, look for alternatives to eliminate them, and make sure you have a plan to manage them appropriately.

Low-Hanging Fruit

____For complex products the ways in which components and final products are packaged and shipped during the "Make" phase can offer opportunities for simple changes that can be both financially and legally beneficial.

____Look for simple ways to cut the weight of your product; this will have positive impacts throughout the rest of the lifecycle.

The "Use" Phase

As we consider the "Use" phase, we've seen the focus turn to the impacts of customer use of the product, or delivery of the service. Energy use is often a main focus of this phase, which brings in legal issues, standards, and the potential to affect the customer's costs. Here's a checklist of key items for the "Use" phase.

"USE" PHASE CHECKLIST

Legal Implications

___Understand all relevant laws and standards regarding energy usage and emissions.

Business Opportunities and Threats

___Look for opportunities to lower the energy costs for your customers through efficiency gains.

___Understand the ease with which energy will be supplied to your product, and make sure it is easy for your customers to obtain and use the needed energy.

___Understand whether there are voluntary energy measures that your customers will expect, and provide them relevant data.

___Look for ways to reduce the customer and environmental impact of product breakdowns and repairs.

Major Impacts

___Use your lifecycle analysis to see whether there are major impacts during the "Use" phase of your product or service that you need to reduce. As in the "Make" phase, these need to be addressed during design time, but also include processes to continue to drive them down over time.

___Look at the mini lifecycles of any consumable used by your product, and understand its impacts and customer implications.

___Remember to look at the "hidden" impacts and the impacts of online services related to your product.

Low-Hanging Fruit

___Look for small changes that can drive your power usage down; small changes may add up and open up different energy delivery opportunities.

The "Renew" Phase

Finally, we've seen that the "Renew" phase focuses on what happens to our products at the end of their useful life. As with the "Make" phase, this is primarily an issue for products. This phase again causes us to look at the materials we use in making the product, as well as how it is put together.

"RENEW" PHASE CHECKLIST

Legal Implications

___Understand all relevant laws and standards regarding product take-back and disposal for the places you intend to sell your product. In many instances, there may be reporting requirements for which you need to have processes in place.

Business Opportunities and Threats

___Try to avoid onerous disposal requirements that will be a burden to your customers.

___Understand what materials you may be able to use that will have value in the recycling market at the end of your product's life.

___Look for ways to simplify the disassembly of your product.

Major Impacts

___Use your lifecycle analysis to see whether there are major impacts during the "Renew" phase of your product that you need to address. These will tend to concern chemicals or materials that are highly toxic and difficult to dispose of responsibly.

Low-Hanging Fruit

___Have your disassembly team or partner look at your product before you commit to manufacture. They may identify small changes that could improve the money you get at end of life, or that could lower the cost to disassemble and recycle your product.

Required Reading for Citizen Engineers

- *Code 2.0,* by Lawrence Lessig (Basic Books, 2006)

- *Cradle to Cradle,* by William McDonough and Michael Braungart (North Point Press, 2002)

- *The Future of Ideas,* by Lawrence Lessig (Vintage, 2002)

- *Green to Gold,* by Daniel Esty and Andrew Winston (Wiley, 2009)

- *Hot, Flat, and Crowded,* by Thomas Friedman (Farrar, Straus and Giroux, 2008)

- *The Mythical Man-Month,* by Frederick Brooks, Jr. (Addison-Wesley, 1995)

- *Remix: Making Art and Commerce Thrive in the Hybrid Economy,* by Lawrence Lessig (Penguin Press, 2008)

- *The Structure of Scientific Revolutions,* by Thomas S. Kuhn, (University of Chicago Press, 1962)

- *Sustainable Energy: Choosing Among Options,* by Jefferson W. Tester, Elisabeth M. Drake, Michael J. Driscoll, Michael W. Golay, and William A. Peters (MIT Press, 2005)

- *Wikinomics: How Mass Collaboration Changes Everything,* by Don Tapscott and Anthony Williams (Portfolio Hardcover, 2006)

- *The World Is Flat,* by Thomas Friedman (Picador, 2007)

- *Zen and the Art of Motorcycle Maintenance,* by Robert Pirsig (Harper Perennial Modern Classics, 2008)

Notes

Preface

1. Sun Microsystems is a signator of the UN Global Compact. For more information see www.unglobalcompact.org/.
2. See http://creativecommons.org/licenses/by-nc-sa/3.0/us/.

Chapter 1

1. Watson, Traci. "For NASA, 'The Right Stuff' Takes on a Softer Tone." *USA Today*, February 4, 2008.

Chapter 2

1. Joy, Bill. "Why the Future Doesn't Need Us." *Wired*, August 2004. For the full article see www.wired.com/wired/archive/8.04/joy_pr.html.
2. Smith, Kirk R. "Wealth, Poverty, and Climate Change." *Medical Journal of Australia* 179: December 2003.
3. Source: "Made in China," by Fara Warner, *Fast Company*, April 2007. To read the full article, see www.fastcompany.com/magazine/114/open_features-made-in-china.html.
4. Source: "2005 CEO Study: A Statistical Snapshot of Leading CEOs," by Spencer Stuart, see http://content.spencerstuart.com/sswebsite/pdf/lib/2005_CEO_Study_JS.pdf.

5. Source: "Pump and Dump Scams Hit Brokerages," by Katie Benner, *Fortune Magazine* reporter; http://money.cnn.com/2007/03/08/markets/pump_dump.fortune/index.htm.

6. Krebs, Brian. "Cyber Criminals and Their Tools Getting Bolder, More Sophisticated." *The Washington Post*, March 14, 2007.

7. Vijayan, Jaikumar. "Hackers Offer Subscription, Support for Their Malware." *Computerworld*, May 4, 2007.

Part II

1. You'll see that we use the words *environmental* and *eco* interchangeably in this book. To some people these are loaded terms; for us there's no particular strategy—or agenda—for using one or the other.

Chapter 3

1. Source: Moisés Naim, editor-in-chief of ForeignPolicy.com; www.foreignpolicy.com/story/cms.php?story_id=4166.

2. Source: "How Much Coal Is Required to Run a 100-Watt Light Bulb 24 Hours a Day for a Year?" HowStuffWorks.com, October 3, 2000; http://science.howstuffworks.com/question481.htm.

Chapter 4

1. Will, George. "Use a Hummer to Crush a Prius." Syndicated April 12, 2007.

2. Visit www.sierraclub.org/sierra/200711/mrgreen_mailbag.asp.

3. North Point Press, 2002.

Chapter 5

1. For more information see http://sic.conversationsnetwork.org/shows/detail3245.html.

2. For more information see the Environmental Defense Fund's paper calculator at www.papercalculator.org.

3. The complete standard is available at www.ghgprotocol.org/standards/corporate-standard.

4. See www.rff.org/rff/RFF_Press/bookdetail.cfm?outputID=8696.

5. As published at Worldchanging.com, May 4, 2006.

6. For more information see www.gabi-software.com/english/gabi/gabi-4/.

Chapter 7

1. National Renewable Energy Laboratory Technical Report, October 2007. For the full report see www.nrel.gov/docs/fy08osti/42266.pdf.

2. Source: Propane Education and Research Council.

3. Source: *Nature's Building Blocks,* by John Emsley (Oxford University Press, 2001), page 479.

4. Source: U.S. Department of Energy; www.fieldstoneenergy.com/pdfs/US%20DepartmentofEnergy.pdf.

5. Source: European Photovoltaic Industry Association; www.epia.org/fileadmin/EPIA_docs/documents.

6. The source of much of the information provided in this subsection is paraphrased from *Sustainable Energy: Choosing Among Options,* by Jefferson Tester, Elisabeth Drake, Michael Driscoll, Michael Golay, and William Peters (MIT Press, 2005). The material is reprinted with permission from MIT Press.

7. A comprehensive inventory of environmental attributes of electric power systems is available at the Emissions & Generation Resource Integrated Database (eGRID) Web site: www.epa.gov/cleanenergy/energy-resources/egrid/index.html. U.S. carbon footprint tables by state are available at www.eredux.com/states/.

8. Chouinard, Yvon. *Let My People Go Surfing: The Education of a Reluctant Businessman* (Penguin Press, 2005).

9. For the full EPA report see www.epa.gov/climatechange/emissions/index.html.

10. Source: Carbon Trust; www.carbontrust.co.uk/resource/measuring_co2/Greenhouse_gas_conversion.htm.

11. Source: www.defra.gov.uk/environment/business/reporting/conversion-factors.htm.

12. See www.epa.gov/cleanenergy/energy-resources/egrid/index.html.

13. See www.epa.gov/climatechange/emissions/ind_calculator.html.

14. For more information about the economic concept of externalities, see http://en.wikipedia.org/wiki/Externality.

15. Published by O'Reilly (2006), available at Safari Books Online.

16. Source: *Power House,* a television program sponsored and produced by Alliant Energy.

17. Source: Justin James, TechRepublic; http://articles.techrepublic.com.com/5100-10878_11-6185330.html.

18. Source: A study conducted by Dr. Jonathan Koomey of AMD, using data from industry analyst firm IDC.

19. Source: IDC, Worldwide Server Power and Cooling Expense 2006–2010 Forecast, Doc #203598, September 2006.

20. Source: The Green Grid; www.thegreengrid.org.

21. Contrarian Minds, by Al Riske, October 11, 2007; http://research.sun.com/.

Chapter 8

1. Source: Environmental Packaging International.

2. European Parliament and Council Directive 94/62/EC of December 20, 1994, on packaging and packaging waste, amended in 2004 and 2005. For a summary see www.epa.gov/swerrims/international/factsheets/200610-packaging-directives.htm.

3. Source: VDI, the Association of German engineers.

4. Published by Raymond Communications, September 2007. For details visit www.raymond.com/library/news/3786-1.html.

5. "It's Waste Not, Want Not at Super Green Subaru Plant," by Chris Woodyard, published in *USA Today*, February 19, 2008.

Chapter 9

1. For additional information see "Virtual Water Trade Between Nations," by Argen Hoekstra, UNESCO-IHE Institute for Water Education, Delft, The Netherlands, at www.gdrc.org/uem/footprints/Hoekstraglobal.pdf.

2. Millennium Ecosystem Assessment Web site—a full range of reports is available here.

3. See http://videogames.yahoo.com/feature/playstation-2-component-incites-african-war/1231745.

Chapter 10

1. For complete details see www.interfacesustainability.com.

2. See www.interfacesustainability.com/coolcarpet.html.

3. See www.interfacesustainability.com/jour.html.

Chapter 11

1. "The Six Sins Of Greenwashing," Terra Choice Environmental Marketing, November 2007.

2. The CorpWatch Greenwash Awards are presented bimonthly to corporations nominated by readers. For details (or the most recent award winners) see www.corpwatch.org/article.php?list=type&type=102.

3. The report is available at www.terrachoice.com/files/6_sins.pdf.

Chapter 12

1. Source: "A New Battlefield: Ownership of Ideas," by Victoria Shannon, Kevin J. O'Brien, and John Markoff, *International Herald Tribune,* October 3, 2005. Please read the whole article—it's excellent.

2. Some argue that patents add teeth to an open source license. That is, if you violate the open source terms we can also prosecute on the basis of patent infringement.

3. The 1-Click patent refers to the idea that on a Web site, a customer can make a purchase with a single mouse click. (For details see U.S. Patent 5,960,411.)

4. Coe, Lewis. *The Telephone and Its Several Inventors: A History* (McFarland, 1995).

5. Source: "Free Culture." Lawrence Lessig Keynote from OSCON 2002, as published in the O'Reilly Policy DevCenter on the O'Reilly Network: www.oreillynet.com/pub/a/policy/2002/08/15/lessig.html.

6. Source: The official Creative Commons Web site; http://wiki.creativecommons.org/Before_Licensing.

7. The definitions and descriptions of *copyleft, FairShare,* and *Share Alike* provided here are extracted from the OSI and GNU Web sites.

8. Source: http://en.wikipedia.org/wiki/Trade_secret.

9. "Your Boss Could Own Your Facebook Profile," The Register, July 16, 2007. See www.theregister.co.uk/2007/07/16/social_networking_profiles_company_property/.

10. You can get the latest information about IP protections in specific countries at www.wto.org/english/tratop_e/TRIPS_e/TRIPS_e.htm.

11. Source: Alicia Beverly, chief IP strategist with IP Wealth, in an article titled "Protecting and Enforcing Your IP Rights in China," October 2006.

12. You can download the report at www.aberdeen.com/summary/report/benchmark/RA_IP_JmB_3676.asp.

13. Source: "Protecting Your Intellectual Property in China," by David McHardy Reid and Simon J. MacKinnon, published in the *Wall Street Journal* March 10, 2008.

14. Source: WIPO Patent Report, Statistics on Worldwide Patent Activity, 2007. The full report is available at www.wipo.int/ipstats/en/statistics/patents/patent_report_2007.html.

15. As published in *Foreign Affairs*, November/December 2002. The full article is available at www.foreignaffairs.org/20021101faessay10075-p10/david-s-evans/who-owns-ideas-the-war-over-global-intellectual-property.html.

16. The full article is available at www.nytimes.com/2008/01/13/business/13stream.html?_r=2&oref=slogin&oref=slogin.

Chapter 13

1. And just to be clear, if you don't own the copyright to some code you almost certainly don't own the right relicense the code.

2. The source of the information and comments about the licenses come from the GNU Foundation. We encourage you to learn more at www.gnu.org/licenses/.

3. At the time of this writing, he doesn't.

4. Read the entire Q&A at http://redmonk.com/sogrady/2008/04/18/ask-redmonk-open-source-indemnification-the-qa/.

Chapter 14

1. "Don't Let Intellectual Property Twist Your Ethos," by Dr. Richard M. Stallman, June 2006.

2. Lessig is the author of *The Future of Ideas* and *Code 2.0,* both of which are required reading for Citizen Engineers.

3. *Research and Development in the Pharmaceutical Industry,* Congressional Budget Office, October 2006.

4. Source: Wikipedia (article on patent history).

5. For details see the article about James Watt at www.wikipedia.org.

6. See Yochai Benkler's *The Wealth of Networks* for details on the relevance to the development of radio (page 191 in the book, or paragraph 350 online).

7. See www.mpegla.com.

8. "Chinese Set-Top Box Makers, MPEG LA Face off Over Patent Fees," *EE Times*, March 12, 2007.

9. Random House, 2001.

10. See http://en.wiktionary.org/wiki/commons.

11. We've heard Stallman quip, "I've never been for open source software. I've been for free software." We confess to conflating the concepts of "open" and "free" here, but to Stallman's point an open license is just one tool in the box.

12. See www.openparenthesis.org/2007/07/25/moglen-oscon.

13. Portfolio Publishers, 2006.

14. See "chocolate.com," *The Economist*, April 17, 2008.

15. See http://openprosthetics.org/.

16. "Open-Source Thinking Revolutionizes Prosthetic Limbs." *Scientific American*, September 17, 2008.

17. See http://news.bbc.co.uk/1/hi/technology/4530930.stm.

18. See www.alexa.com/.

19. See http://en.wikipedia.org/wiki/Wikipedia, which is a deliciously circular reference!

20. The leading examples of which are VHDL and Verilog.

Chapter 15

1. From an interview in *Linux Electronics*; www.linuxelectrons.com/article.php/20060322062359676.

2. Sun announced the Open Media Commons (OMC; www.openmediacommons.org) initiative in August 2005 as an open source community project to develop royalty-free open solutions for digital content, including DRM solutions.

3. Project DReaM began as an internal Sun research effort and then transitioned to a community project with the announcement of the Open Media Commons. The goal of Project DReaM is to encourage community participation in the development of CAS and DRM/"Mother May I" (DRM-MMI) specifications and open source reference implementations based on Sun's initial contributions from Project DReaM. The specifications were initially drafted by Sun and made available to interested parties who registered with OMC. Simultaneously, open source reference implementations by Sun were made available under Sun's Common Development and Distribution License (CDDL).

Chapter 16

1. As published in *Engineering Enterprise,* the alumni magazine for the Georgia Institute of Technology, Spring 2004. To read the full article see www.lionhrtpub.com/ee/spring04/nature.html.

2. Source: National Academy of Engineering (NAE), "Educating the Engineer of 2020"; www.nap.edu/openbook.php?isbn=0309096499.

3. "A Call for K-16 Engineering Education," by Jacqueline F. Sullivan, *National Academy of Engineering Publications* 36(2): Summer 2006: www.nae.edu/NAE/bridgecom.nsf/weblinks/MKEZ-6QDLB3?OpenDocument.

4. For more information see www.cee.cornell.edu/?page_id=21.

5. See www.kettering.edu for additional information.

6. For more information see www.cals.vt.edu/students/undergraduate/majors/bse.html.

7. See ENG-1001; www.geneng.mtu.edu/courses.html.

8. For additional details see http://registrar.utexas.edu/catalogs/ug08-10/ch09/index.html.

9. Ms. Chang and her team at Sun Labs were instrumental in developing ECC, the cryptographic technology underlying the two dominant security libraries (OpenSSL and Mozilla/NSS).

Chapter 17

1. This article is based on a talk given October 10, 2005, at the NAE Annual Meeting.

2. The complete article is available at www.ewb-international.org/pdf/TimesofIndia.pdf.

3. See the EWB Web site for further details: www.ewb-international.org/.

4. For more information see www.berkeley.edu/news/media/releases/2006/06/06_telemedicine.shtml and http://tier.cs.berkeley.edu/docs/wireless/large_wild.pdf.

5. For details see www.playpumps.org.

6. These are really cool. For details check out www.sunspotworld.com/.

7. You can read the full article at www.earthinstitute.columbia.edu/news/2007/story04-24-07.php.

8. Margolis, Mac. "Bringing Spice to a Hot Land." *Newsweek,* December 10, 2007.

9. To read the complete article visit www.newsroom.ucr.edu/cgi-bin/display.cgi?id=1719.

Photo Credits

INDEX

FREE Online Edition

Your purchase of **Citizen Engineer** includes access to a free online edition for 45 days through the Safari Books Online subscription service. Nearly every Prentice Hall book is available online through Safari Books Online, along with more than 5,000 other technical books and videos from publishers such as Addison-Wesley Professional, Cisco Press, Exam Cram, IBM Press, O'Reilly, Que, and Sams.

SAFARI BOOKS ONLINE allows you to search for a specific answer, cut and paste code, download chapters, and stay current with emerging technologies.

Activate your FREE Online Edition at
www.informit.com/safarifree

> **STEP 1:** Enter the coupon code: SGFKYFA.

> **STEP 2:** New Safari users, complete the brief registration form.
> Safari subscribers, just log in.

If you have difficulty registering on Safari or accessing the online edition, please e-mail customer-service@safaribooksonline.com